Advance Praise for *Virtual Teams*

"... very well-written, clearly bringing forth the early essence of what I think is the most important foreseeable potential of the erupting digital technology—i.e., enabling rapid evolution of organizational forms representing much-improved species of social organisms. Read it; I think human survival depends upon this."

　　—*Douglas Engelbart, Director of the Bootstrap Institute*

"*Virtual Teams* emphasizes the practical how-to's of interpersonal bonding that allow industry/academic partnerships to utilize the technology of the network to excel as virtual teams."

　　—*Bill Hanson, Industry Co-Director, MIT's Leaders for Manufacturing Program*

"Lipnack and Stamps have been thinking about teams and networks longer and writing about them more than most. For those who want to lead the movement, catch up with it, or simply know where it is going, their books are packed with useful information and interesting stories."

　　—*Dee W. Hock, Founder and Chairman Emeritus, VISA*

"This book provides a long overdue perspective on how to apply the discipline of real teams in the fast moving, increasingly dispersed information age of the future."

　　—*Jon R. Katzenbach, Co-Author of* The Wisdom of Teams

"*Virtual Teams* provides valuable insights into global teamwork and management through network technologies now available to all companies, large or small."

　　—*Jim Lynch, Director, Corporate Quality, Sun Microsystems, Inc.*

"Lipnack and Stamps have written an important handbook for the 21st century corporation. It is a practical guide for using the new tools of networking technology to leverage the productivity, efficiency, and genius of teams. For virtual teams, time and space constraints vanish without its members sacrificing their individual contribution, enthusiasm, and interest. Here you will read about the dynamic organizations of the future."

　　—*Regis McKenna, The McKenna Group, Author of* Relationship Marketing

"This book fills a big gap in what's available to help teams. Jessica and Jeffrey go beyond the theory on networking people and teams to actual how-to and tools to get the job done. When you are working on multiple distributed projects, you need a reliable process and tool kit."

　　—*William Miller, Vice President, Research and Business Development, Steelcase, Inc.*

"*Virtual Teams* is chockful of striking examples of how technology and innovative people are reshaping the way companies compete globally."
—*Lars Nyberg, Chairman & CEO, NCR Corporation*

"Lipnack and Stamps have done it again. Having shown the world long before it became obvious how to use networks of smaller independent companies to outperform bureaucratic giants, they now show us how and when to use virtual teams to overcome the barriers of location, time, and organizational separation. Essential lessons for anyone seeking to be effective in the information age."
—*Gifford and Elizabeth Pinchot, Authors of* The Intelligent Organization and The End of Bureaucracy

"Lipnack and Stamps wrote the original book on networking years ago. While others seem to have recently discovered the possibilities inherent in the convergence of new ways of organizing teams and new communication technologies, they have been working at the bleeding edge for decades. If you want to see where organizational communications are going in the future, heed what these pioneers have written today."
—*Howard Rheingold, Author of* The Virtual Community *and Founder of* Electric Mind

"*Virtual Teams* is *the* cookbook for the emerging network-based organization. The role models portrayed clearly demonstrate just what can be accomplished when we break down the traditional barriers of space and time. Even already practicing virtual team members will learn new 'tricks' from this marvelous compendium, driving their organization towards even more effective ways of working."
—*Harry J. Saal, Smart Valley, Inc.*

"Jessica and Jeffrey are right on the mark! I have participated on virtual teams for over a decade with the likes of Adobe Systems, Apple Computer, and Oracle Corp. And now with Borland International I interact daily with various teams whose participants are in many functional areas located around the world. A lot of these teams are 'Intra'-Borland, but many are 'Inter'-Borland in that they include participants from other companies, accounting, and consulting firms."
—*Del Yocam, Chairman and CEO, Borland International, Inc.*

VIRTUAL
TEAMS

Books by Jessica Lipnack and Jeffrey Stamps

Other Books in This Trilogy

The Age of the Network: Organizing Principles for the 21st Century

The TeamNet Factor: Bringing the Power of Boundary Crossing into the Heart of Your Business

Other Books by the Authors

The Networking Book: People Connecting with People

Networking: The First Report and Directory

Holonomy: A Human Systems Theory *by Jeffrey Stamps*

VIRTUAL TEAMS

*Reaching Across Space, Time, and
Organizations with Technology*

JESSICA LIPNACK & JEFFREY STAMPS

JOHN WILEY & SONS, INC.

New York • Chichester • Weinheim • Brisbane • Singapore • Toronto

ACKNOWLEDGMENTS

A book is a complex project. We have been blessed with the help of many people who are our colleagues in trying to understand the new organizational phenomenon of virtual teams.

Jim Childs, our editor and publisher on our last two books and our publisher once again on this one, was the first to express interest in the book. This long collaboration continues to inspire us. Jeanne Glasser, our editor, has been an enthusiastic source of energy and encouragement. Publishing with the nearly two-centuries-old John Wiley & Sons is an honor.

Reuben Harris, chair of the Department of Systems Management at the Postgraduate Naval Academy and a dear friend, gave us early and invaluable guidance on how to approach this topic. Bob Buckman of Buckman Laboratories calls Reuben "America's best-kept business secret." Perhaps we can help put an end to that secret.

A group of devoted people took the time to line edit the manuscript for no reason other than that the topic interests them. They are the kinds of friends everyone wants. Their hard-hitting queries caused us to write a better book: Kathi Albertini, Frank Alla, Rick Berenson, Don Brown, Curt Crosby, Bernie DeKoven, David Paeper, Pete Rogers, Caryn Siegel, Gary Wheeler, and Peter Wiley. This was the second time Kathi and Frank provided this service to us. And Frank did triple duty—we also interviewed him for the book, as we did Curt, Bernie, and Gary. David and Caryn generously volunteered from afar. Pete, a member of our extended family, did such an extraordinary job as an editor that he may have yet another career in his future. Peter went far beyond the call of duty. Abiding thanks to all of you.

Cathy and Ron Cordes, colleagues and dear friends, were unwavering in one commitment. Weekend after weekend during the book's writing, they made sure that we ate at least one very delicious meal. They also made us laugh. We use their wonderful product TeamFlow in all our projects. Judy Smith, another extended family member, called innumerable times to ask how it was going, as she has on all our books. For nearly 20 years, she has housed us when we visit our publisher.

As part of our research for this book, we interviewed 75 people. We thank them for the time they took to carefully consider the complex questions that face organizations when people work across boundaries:

Rebecca Adamson, First Nations Development Institute;
Frank Alla, Management Insight Technologies;
Haluk Ariturk, The Acacia Group;
Fern Bachman, Jr., Apple Computer;
Victor Baillargeon;
Larry Banks, Hewlett-Packard;
Erik Boe, Apple Computer;
Antonia Bowring, Women's World Banking;
C. Marlin Brown, Sun Microsystems Computer Company;
Don Brown, NASA/Johnson Space Center;
Harry Brown, EBC Industries, Inc.;
Bob Buckman, Buckman Laboratories;
Peter Buesseler, Minnesota Department of Natural Resources;
Rita Cleary, The Learning Circle;
Camen Criswell;
Curt Crosby, Sun Microsystems Computer Company;
Bill Crowley, SunExpress;
Richard Dale;
Earnest W. Deavenport, Eastman Chemical Company;
Bernie DeKoven, Mattel Media;
Timm Esque, Intel Corporation;
Susen Fagrelius, Susen Fagrelius & Associates;
Mike Fraser, National Oceanographic and Aeronautics Administration;
George Gates, Core-R.O.I.;
David Gibson, SunService;
Al Gilman;

Diane Gilman, Context Institute;
Loree Goffigon, Gensler;
Rogier Gregoire;
Georgene Hanson-Witmer, Boise Cascade;
Nadia Hijab, United Nations Development Programme;
Will Hutsell, Eastman Chemical Company;
Jean Ichbiah, Textware Solutions;
Sture Karlsson, Tetra Pak Converting Technologies;
Rob Lau, Steelcase, Inc.;
John E.S. Lawrence, United Nations Development Programme;
Celestine Lee, SunService;
Ed Lynch, Identity Clark County;
Jim Lynch, Sun Microsystems;
Kathleen Shaver Madden, Foundation for International Nonlinear
    Dynamics;
Pamela Martin, Apple Computer;
Ginger Metcalf, Identity Clark County;
Arch Miller, Executive Forum;
Jeff Morgan, *Men's Health;*
Charles T. Nason, The Acacia Group;
Ike Nassi;
Dan Nielsen, Voluntary Hospitals of America;
Aron Oglesby, Dance New England;
Lynda Popwell, Eastman Chemical Company;
Leslie Rae, Science Applications International Corporation;
Frank Reece, US TeleCenters;
Dennis Roberson, NCR Corporation;
Charlie Robertson, Red Spider;
Judi Rosen, CSC Index;
Rustum Roy, Pennsylvania State University;
Harry Saal, Smart Valley, Inc.;
Scott Simmerman, Performance Management Company;
Nova Spivack, EarthWeb;
John Steele, Eastman Chemical Company;
Brian Stenquist, Minnesota Department of Natural Resources;
Evelyne Steward, The Calvert Group;
Carie Strahorn, Boise Cascade;
W.R. Sutherland, Sun Microsystems;

Steve Teicher, Apple Computer;
Steve Tromp, First Community Bank/BankBoston;
Alison Tucker, Buckman Laboratories;
Valerie Veterie, Pfizer, Inc.;
Jim Vezina, Carnegie Group, Inc.;
Art Wagner, Blue Sky Advertising/Marketing, Inc.;
Irvenia Waters, APM, Inc.;
Linda J. Welsh, Sun Microsystems;
Clay Wescott, United Nations Development Programme;
Gary Wheeler, Perkins & Will/Wheeler;
Scott Woods, SunService;
Yurij Wowczuk, Wheeling Jesuit University;
Terri Yearwood, Minnesota Department of Natural Resources; and
Tom Young, SunService.

Another group provided help in connecting us with our interviewees: Calvin Colbert, Voluntary Hospitals of America; Rod Cox and Lynn Swartzlander, Executive Forum; George Brennan, Bob Farkas, Debbie Moore, and Lola Signom, NCR Corporation; Patricia Fiske, Worldwide Partners; Alex Kleinman, The Advisory Board Company; Bahman Kia and Jose Cruz-Osorio, United Nations Development Programme; Barbara McConnville, Buckman Laboratories; Tony Pribyl, *Men's Health;* and Patricia Sutton and Nancy Ledford, Eastman Chemical Company.

Annie Marascia, who has lived through three books with us, has been tireless—a sophomore in high school when she started with us, she is now a sophomore in college. Lorienne Schwenk has been a talented research assistant.

Nancy Marcus Land at Publications Development Company of Texas has made this book's production a virtual team dream.

Marion Metcalf and Lisa Kimball provided key insights as they both did on our last book. Thank you, sisters. Al Gilman was an early and perceptive contributor to the ideas.

Our clients and their organizations are where we learn the most. We are especially grateful to Steve Teicher and Pam Martin at Apple for many hours spent on various continents exploring virtual teams in practice. Gail Snowden of First Community Bank/BankBoston and Heather

Stoneback of Rodale Press have taught us things we could only know by learning about their organizations. We are also grateful to our many friends at Steelcase, including Bill Miller, Doug Parker, and Kyle Williams. Steelcase makes it possible for virtual teams to work together. Our 15-year connection with The Calvert Group has taught us how values sustain virtual teams. Our colleagues with whom we organized "MassNet: Collaboration for the CommonWealth" showed our local community what can happen when diverse organizations decide to cooperate.

Finally, as has been the case with all our books, our daughters, Liza and Miranda, are the point of it all. Having lived with these ideas since they were born, they know more than anyone.

JESSICA LIPNACK
JEFFREY STAMPS

West Newton, Massachusetts
February 1997

CONTENTS

LIST OF ILLUSTRATIONS

INTRODUCTION

COMING HOME

Virtual Teams is the final book in a trilogy on network organizations that we have been writing since 1991. With this book, we bring networks down to earth, to people who work with others in small groups stretched across space, time, and organizations. Having asked the question of hundreds of very diverse audiences, we know that most of you regularly work with people located more than 50 feet from your workplace. Therefore you have at least a distance problem to solve in order to work collaboratively (see Chapter 1).

Virtual Teams is about a radically new type of small group emerging in the Age of the Network. These boundary-spanning centers of people-to-people activity are the social cells that make up larger network organizations. They are small task-oriented groups from the executive suite to the front line. Webs of interactions and relationships bind them together.

- **Teams:** In *Virtual Teams*, we take a deep look at how this most fundamental organization—the team—is transforming ("morphing," in computer lingo) into an extraordinary new 21st-century version. We focus on small groups of people working across boundaries supported by the new computer and communications technologies. Increasingly, this is the reality of everyday work life for many people.

- **Teamnets:** In our 1993 book, *The TeamNet Factor,*[1] we center on the network as a form of organization. We show its variations at every size from small groups, to enterprises, to alliances, to nations. In that book, we coin the word *teamnet* to put people back into networks and to emphasize their multilevel (groups within groups) nature. We show how networks offer practical approaches to solving old problems and launching new initiatives. We also offer three chapters on methods to develop networks along with several chapters that focus specifically on small business networks.

A concept for supply chain co-op

- **Networks:** In our 1994 book, *The Age of the Network,*[2] we provide an overview of the impact of networks and their strategic importance. There, we place networks—the signature organization of the Information Age—in the context of bureaucracy, hierarchy, and small groups, which dominated earlier eras. We show how companies use networks to their strategic advantage. These nimble, boundary-crossing configurations also incorporate what is uniquely valuable about each of the earlier forms.

In the years since we began writing this trilogy, technologies that directly impact networks have significantly expanded the spectrum of options for people to connect with one another. One noticeable example is this now ubiquitous form of address that we see many times each day:

Visit our Web site at http://www.netage.com

Cyberspace words, known only by a select group in the early 1990s, have become daily occurrences—such as the Internet, the World Wide Web, hypertext, and intranets. All these and many more technologies contribute to a dramatically extended ability for teams to work together at a distance.

This book is about organizations that spread out and reach across boundaries. They do so with the help of and in response to technology. Technology extends our capabilities, but organizing to do things together is still a human capability. The people side of the organization/technology relationship is the focus of *Virtual Teams*.

Experience at the Source

Our research combines the knowledge of the people we interviewed with our own experience over many years working within and between organizations as consultants and participants.

With *Virtual Teams*, we come home to the heart of our personal experience. We have always worked in small groups across distances and organizational boundaries.

Research for this book began long ago at the dawn of our relationship in 1968, when we met as students at Oxford University. A few years later, married, living in our first (and still only) house with our first "personal computer" (a Wang 600 Programmable Calculator), we began life as independent entrepreneurs with a consulting business.

Virtual teams have been a way of life for us for twenty-five years. We have partnered with thousands of people on a wide range of projects for clients in every sector—from Digital Equipment Corporation, to the national Presbyterian Church, to the U.S. Department of Commerce.

Since 1979 when we began to contact people and gather information for our first book on networked organizations (*Networking*,[3] published in 1982), we have received volumes of material from all around the world. We have heard from people in more than half the world's countries and visited with networkers from every continent, including Antarctica!

As writers, researchers, speakers, seminar leaders, and consultants, we have known and been part of many very different types of organizations. From engagements that lasted only a few hours, to projects of a few days, to multiyear programs, we have acted as "drop-in" outside experts, involved facilitators, core members, and leaders of customer teams. We have worn corporate badges, received passwords to internal computer systems, and occupied offices within our clients' buildings. We have even worn the badges of our customers' customers. Although we draw primarily on examples from the business community in this trilogy, our quarter-century of research and experience is also extensive in government, nonprofit, and grassroots organizations. These are the focus of our first two books, *Networking*, and its 1986 successor, *The Networking Book*.[4]

Theory and Application

Since our first book on networks, we have strived to integrate our work into a coherent conceptual framework supported by general systems theory. Systems theory is about principles and patterns of organization that apply across scientific disciplines—notoriously difficult boundaries to cross. Human systems were the subject of Jeff's 1980 book (and doctoral dissertation), *Holonomy*.[5] Systems principles have helped us recognize common patterns among the awesome variety of human organizations, particularly the core features of the newly emerging forms. They provide a powerful infrastructure for the network organizational model that we have been developing and testing for almost two decades.

Theory is very practical. It enables quick adaptation of shared learning to always unique circumstances. Theory provides a consistent, shareable, knowledge-based approach to develop and manage virtual teams. Theory also provides a framework to test ideas and improve practical knowledge about how to work collaboratively at a distance.

Originally, we recognized ten principles[6] of network structure and process, which threaded through our first two books. We consolidated the principles to five[7] in *The TeamNet Factor* and *The Age of the Network*. Since frontier knowledge never stands still, we reconsidered the principles for this book. In our ongoing effort to improve our conceptual tools, we have:

- Simplified the basic elements of a virtual team to three—people, purpose, and links; and

- Expanded the principles from five to nine, which provide a more comprehensive set of guidelines for the "care and feeding" of virtual teams.

When Things Go Awry

Virtual teams are not a panacea for teams that do not work. Quite the contrary. It is harder for virtual teams to be successful than for traditional face-to-face teams. Misunderstandings are more likely to arise and more things are likely to go wrong.

We are not cheerleading for this gee-whiz-it's-a-new-and-better-way-to-do-things approach. Rather, our goal is to understand and improve virtual teams. Virtual teams are already prevalent and increasingly more will appear in the years ahead. Indeed, lack of recognition that teams have gone virtual contributes to the high failure rate of today's teams. When teams spread out, they have a dramatic effect on the entire management structure. There are more virtual teams working at all levels than you realize, and the way they work is likely different from what you think.

Everything that goes wrong with in-the-same-place teams also plagues virtual teams—often it is worse. Egos, power plays, backstabbing, hurt feelings, low confidence, poor self-esteem, leaderlessness, and lack of trust all weaken virtual teams. When communication breaks down, it requires that people take measures to repair it. It is just that much more difficult to communicate across distance and organizations.

Many of the problems that teams encounter are ancient in nature. Millennia of face-to-face exchanges inform most of our collective experience, tools, techniques, and lore. Methods that work to correct problems that arise in face-to-face teams are only a starting point for virtual teams.

We address the problems of *virtual* teams as directly as possible and present what people do to solve them. At the same time, we encourage you to draw on what you already know about teams. For example, what do you do if a virtual team member is not participating? The same thing you do if a face-to-face team member is not participating. Talk to that person by any means possible, find out what is preventing participation, and solve the problem.

Common sense and the large body of excellent material developed by team experts over the past several decades provide some solutions to these problems. We reference these sources extensively in the Notes. Gradually, a body of detailed knowledge and technique will develop for the field-in-the-making, "Virtual OD" (Organizational Development).

To Be Expert In This !

We do not go into detail about *why* companies form virtual teams. So far as we can tell, companies create these distributed organizations for myriad reasons. People form virtual teams when things go wrong, when the people required to do a project happen to be spread out, and just because virtual teams are the best way to get things done in a particular instance.

Our purpose here is to present excellent examples of virtual teams and our thinking about how virtual teams *can* work and meet challenges. Thus, this is a book that shares best practices, not one that critically examines corporate behavior. In time, as the body of information grows, critical analysis will be essential to secure the foundations of network knowledge.

Finally, we are still in the early phases of the transformation from hierarchy-bureaucracy to networks. Virtual teams will expand as a key way to work for as far as we can see into the future. Consider this book an opening view of a future society of work.

Options for Reading This Book

Different people have different preferred ways of learning new information. Some learn best from stories based on experience, some prefer theoretical approaches, others need practical ideas, and most of us need some vision to motivate us to move to the new ways of working. Readers of our previous books will recognize how we paint a whole canvas from these four kinds of views on this difficult-to-grasp subject.

- Vision (insight);
- Stories (experience);

- Principles (theory); and
- How-to (practice).

While we have written a traditional book crafted with loving care to flow from beginning-to-end, we know that people have their own styles for reading books. Some of you begin at the end, some in the middle, while still others skim to find something of interest.

- For an introduction to virtual teams, a definition, some examples of how companies use them in both low and high tech "versions," and an overview of the principles, read Chapter 1, "Why Virtual Teams?"
- For people who learn primarily through the stories of how others have done it, turn to the opening sections of Chapters 2 through 7. There you will find the detailed case studies of six companies with impressive and sometimes astonishing virtual teams. Many other examples appear throughout the text.
- For those who prefer concepts and models, read the sections at the end of those chapters. There you will find an integrated framework to understand and manage this new form of organization. We include important contributions from other writers and researchers.
- For those most interested in how these ideas apply as tools and methods, go to Chapter 8, "Working Smart: A Web Book for Virtual Teams," our handbook for starting and maintaining virtual teams.
- For those who wish to begin with a vision of what virtual teams mean for society, turn to Chapter 9, "Virtual Values."

You can gain a quick overview of the book by reading the headings, looking at the illustrations, and noting the emphasized phrases:

Key ideas in the book are in pull-out quotes that look like this.

As in our previous books, we provide extensive Notes so that readers can go directly to our sources and learn more for themselves. The abundance of material available through the World Wide Web made it easy for us to track down many facts and locate specific sources. We include the addresses for the Web sites that we reference in the book. See our Web site for more detailed information and practical pointers on virtual teams.

CHAPTER 1

WHY VIRTUAL TEAMS?

The New Way to Work

The conventional way in which people work is coming unglued.

Until recently, when you said that you worked with someone, you meant by implication that you worked in the same place for the same organization. Suddenly, in the blink of an evolutionary eye, people no longer must be in the same place—*collocated*[1]—in order to work together. Now many people work in *virtual teams* that transcend distance, time zones, and organizational boundaries.

Today's trend is tomorrow's reality: In the coming decades, most people will work in virtual teams for at least some part of their jobs.

Human beings have always functioned in face-to-face groups. While the use of teams is on the rise—the *Wall Street Journal* reports that two-thirds of American companies employ them—the face-to-face aspect of normal working relationships is changing dramatically.[2] Electronic communication and digital technologies give people an historically unprecedented ability to work together at a distance. Now there is a powerful trend to team across organizational boundaries.

1

Today, people frequently work across internal boundaries—the specialized functions and divisions within their companies. And they often work across external boundaries—in partnership across corporate lines with vendors and customers, in alliances with complementary enterprises, and even in association with direct competitors. A new form of boundary-crossing team is emerging as the basic working unit of the Information Age organization.

BLURRING DISTINCTIONS

Virtual teams are the peopleware for the 21st century.

The onrushing explosion in information and communication technologies makes change in how we team inevitable:

- Dataquest, the technology market research firm, predicts that personal computer (PC) sales, of which there were none in the 1960s, will top 100 million annually by the year 2000[3]—one PC for every 60 people on the planet; and
- By the same time, according to Action Cellular Network,[4] more than 60 million people will use cellular phones—which did not exist in the 1970s.
- Voicemail, rare in the 1980s, is now widespread and all but indispensable in most organizations today.
- Fastest growing of all in the 1990s is the Internet and the World Wide Web, with its internal offspring, intranets. The number of new Internet connections each day surpasses anyone's ability to accurately count them. According to Matrix Information and Directory Services, which has tracked Internet growth for years, electronic connections among people and computers are expanding perhaps on the order of 100 percent annually.[5]

Distance-spanning communication tools open up vast new fertile territory for "working together apart."[6] For the first time since nomads moved into towns, work is diffusing rather than concentrating as we move from predominately industrial to informational products and services.

In all industries and sectors, people are working across space and time. Virtual teams thrive in big companies like Hewlett-Packard and Eastman Chemical Company, in smaller ones like Rodale Press and Buckman Laboratories, and even smaller ones known only to their own markets like Tetra Pak Converting Technologies and US TeleCenters. In government agencies large and small, such as the U.S. Department of Commerce and Minnesota's Department of Natural Resources, in education including Maine's Center for Educational Services and the Massachusetts Teachers Association, and in nonprofits like New York-based Women's World Banking and Boston-based Dance New England, small groups of people work together across boundaries.

How do these new virtual teams form? Sometimes a sudden need to work cross-organizationally sparks their formation. Such was the case recently in the magazine industry.

Like a Rolling Stone

With a circulation of 1.3 million, *Men's Health,* whose moniker is "tons of useful stuff for regular guys," is the second largest men's magazine in the United States. It has in a few short years outstripped its two biggest competitors—*Esquire* and *Rolling Stone*—both of which have been in business decades longer. (*Men's Health* has grown so rapidly that it has surprised even its publisher Rodale Press, famous for its flagship magazines, *Prevention* and *Organic Gardening.*) The combined circulation of the three men's magazines just about equals that of *Sports Illustrated,* the biggest men's publication in the United States with three million subscribers.

Suddenly in 1995, the three smaller arch competitors found themselves working on a crash project as partners in a virtual team brought together by a mutual client.

"We compete with *Rolling Stone* and *Esquire* for the same advertising business," explains Rodale vice president and *Men's Health* publisher Jeff Morgan. "One day Goodby Silverstein [the San Francisco advertising agency] that represents our client, Haggar [a men's clothing manufacturer], came to *Rolling Stone, Esquire,* and us with a challenge. Haggar would either buy its new ad campaign from the three of us together or from a combination of titles within Time Inc., including *Sports Illustrated.* All of us were dumbfounded. We're the biggest competitors there are and we had literally a week to become partners!"

It was not a small sale. It meant multiple pages of advertising that would run over the next two years in each magazine. It offered direct exposure to Haggar's market in the retail environment: The three competitors would jointly custom publish a 14-page guide for casual fashion. The 300,000-piece press run would go to customers when they bought Haggar products. It also would become a give-away in in-flight magazines and in health clubs around the United States. *Men's Health* regarded it as "the Cadillac of value-added projects."

A year later you could still hear the excitement—and the outcome—in Morgan's voice: "The client wound up choosing the three of us rather than Time Inc. This was an important win for us and it's the first time I've ever heard of this in publishing."

How did they do it? They combined face-to-face meetings with telephone conference calls and many faxes. Low-tech by today's standards but electronic nonetheless.

People from each magazine's Bay Area office attended the first meeting in San Francisco, home to the advertising agency Goodby Silverstein & Partners. Then a conference call took place with 15 people on the phone at the same time—from San Francisco, Chicago, Dallas, and New York where all the magazines' advertising directors and publishers are based.

This initial brainstorming session generated enough ideas for the New York contingents of the three publications to take over. They in turn got together at *Rolling Stone*'s offices where each of the magazine's marketing directors threw their ideas on the table. Together, they fleshed out the basic concept. A week later the advertising directors of the three magazines flew to Dallas and presented their idea to the clients, Haggar and Goodby Silverstein. Next the advertising agency worked over the phone, coaching the virtual team on how to present the ultimate look. Finally, within three short weeks of the very first phone call, Haggar selected the joint proposal of the virtual team of competitors over that of the media giant.

"This never would have happened if Goodby Silverstein hadn't proposed it," Morgan says. "But it was such a unique proposition that we just got our heads in there, did it, and then walked away. This is the wave of the future." Considered a harbinger of things to come, the project was written up in the advertising trade press as an example for others to follow.

In record time, the three magazines created a business-winning virtual team. Had they not done so, all would have lost. As with many other industries, losing is not an option for the rapidly changing, highly competitive world of publishing.

The *Men's Health* story may seem a bit extreme—working at light speed with your arch competitors to succeed at jointly winning a highly lucrative contract. It requires an attitudinal shift that traditional business practice does not support. But it is no longer all that exceptional. Teams of people working across boundaries of space, time, and organizations are increasingly common. What is so new is the easy availability of technology to make it happen. Even the Haggar ad campaign proposal could not have succeeded in such record time without the widespread use of communication technology. Tools such as easy-to-set-up conference calls, PCs, and fax machines everywhere were in limited use even as recently as a decade ago.

The Virtue of Virtual

It was not until the 1990s that the word "virtual" made it into the headlines on a regular basis. As a word, virtual has the same Latin root as virtue, an intimately personal quality of goodness and power. Its archaic meaning is "effective because of certain inherent virtues or powers," an apt expression for successful virtual teams.

More recent use brings newer meanings:

- Virtual as in "not in actual fact" but "in essence," "almost like"; and
- Virtual as in "virtual reality."

The "almost like" part of the definition, as in "they act virtually like a team," is on target. "Virtual" is used in the same way in the terms "virtual corporation," "virtual organization," and "virtual office." A virtual team conjures up a different picture from the one of people in the same organization working together in the same place.

When we use the term virtual, we do *not* mean it as another dictionary definition puts it: something that is "not real" but "appears to exist," something "that appears real to the senses" but is not in fact. It is a bit like the old TV commercial about a brand of audiotape: "Is it live or is it Memorex? With Memorex, you can hardly tell."

With a virtual team, can you tell? It feels like a team and acts like a team but is it a live team? Answer:

> *Virtual teams are live, not Memorex. They are most definitely teams, not electronic representations of the real thing.*

The newest meaning of "virtual" attests to forces that are fast moving teams into an altogether different realm of existence—virtual reality—or more precisely, *digital* reality. Electronic media together with computers enable the creation of spaces that are real to the groups that inhabit them yet are not the same as physical places. The eruption of the World Wide Web in the last decade of the millennium has allowed virtual teams to create private electronic homes. These interactive intranets—protected members-only islands within the Internet—signal a sharp up-tick in the human capability to function in teams.

> *Virtual teams are going digital, using the Internet and intranets.*

And the Definition Is . . .

So what exactly is a virtual team? A virtual team, like every team, is a group of people who interact through interdependent tasks guided by common purpose.

> *Unlike conventional teams, a virtual team works across space, time, and organizational boundaries with links strengthened by webs of communication technologies.*

The image of face-to-face interactions among people from the same organization typifies our older models of teamwork. What sets virtual teams apart is that they routinely *cross boundaries*. What makes virtual teams historically new is the awesome array of interactive technologies at their disposal. Virtual teams now use myriad electronic technologies to cope with the opportunities and challenges of cross-boundary work.

Regular meetings, encounters in the hallway, getting together for lunch, dropping into one another's offices—these are our standard methods for getting things done. They lag behind everyday reality. People rarely see one another when they are in different places, spread out around the world, or even housed in different parts of the same city. Motorola, for example, has some 20 locations just in the Northwest Chicago area, each of which has multiple buildings. In the most extreme cases, some teams *never* meet face-to-face but work together online. Such is the case with the 1200 employees of Buckman Laboratories in Memphis, Tennessee, who form and disband numerous situation-specific virtual teams on a daily basis—even though they are spread all around the globe.

A major reason that many of today's teams are ineffective is that they overlook the implications of the obvious. People do not make accommodation for how different it really is when they and their colleagues no longer work face-to-face. Teams fail when they do not adjust to this new reality.

Close Is Really Close

What first comes to mind when you think of a team? A group of people working side-by-side, in close proximity to one another—a basketball or a rugby team, perhaps.

How close do you have to be to get the advantage of being in the same place? That is, what is the "radius of collaborative collocation?" The startling data that MIT Professor Tom Allen has been compiling for the past several decades show that the radius is very small.

> *Based on proximity, people are not likely to collaborate very often if they are more than 50 feet apart.*[7]

The probability of people communicating or collaborating more than once a week drops off dramatically if they are more than the width of a basketball court apart. To get the benefit of working in the same place, people need to be quite close together.

To put this in perspective, think of the people you regularly work with. Are they all within 50 feet of you? Or are some of your coworkers a bit more spread out, down the hall, on another floor, in another building, or perhaps in another city or country? Increasingly, the people we work with routinely are no longer within shouting distance. Any team of more than about 10 to 15 people is by sheer physical mass probably more than 50 feet apart (Figure 1.1).

From a team perspective, the important distances are the personal ones. How close people like to be for interpersonal interactions varies by culture.[8] How far away do people have to be before they need to worry about compensating for distance?

The farther apart people are physically, the more time zones they have to cross to communicate. Thus, time becomes a problem when people who are not in the same place need some of their activities to be in sync. The window for routine synchronous work shrinks as more time zones are crossed, closing to effectively zero when people are on opposite sides of the globe. People who work together in the same place also can have time problems. Salespeople or consultants, for example, rarely occupy their offices at the same time. Even apparently collocated teams often cross time boundaries and need to think virtually.

My Organization Is Your Organization

Do all the people you work with to get your job done work for the same organization? Probably not.

Figure 1.1 Collocated to Virtual Distance

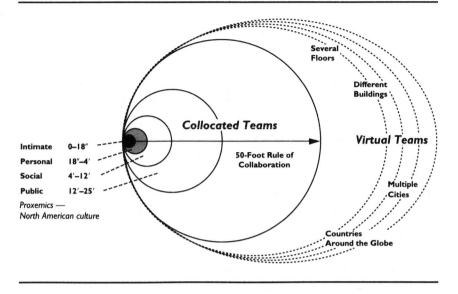

Most core business processes require that people regularly work across organizational boundaries. Supply chain management, marketing, product development, sales, quality improvement, and change management are just a handful of activities that require virtual teams to work over walls and across borders.

Large-scale systems change invariably requires teamwork across organizational borders. To reinvent its administration and information management system, the U.S. Department of Commerce has involved hundreds of people in teams from five major bureaus and dozens of smaller organizations. Usually numbering 8 to 10 people each, these virtual teams also involve scores of contractors who provide everything from consultation on change management to software programming.

When Acacia Mutual Insurance Company in Washington, DC, decided to have a third-party administrator do the processing of its new variable universal life insurance product, it immediately created a virtual team with its supplier, Financial Administration Systems, located in

Connecticut. Alliances, joint ventures, and partnerships all require companies to establish cross-boundary teams of small groups from member organizations.

Acacia's ability to easily team with a third party draws on its decade-long change effort that began when a new CEO arrived in 1988. "I wanted people to embrace customer service and have a team orientation," says Charles T. ("Tuck") Nason, also Acacia's chairman. "It was a very bureaucratic, function-oriented culture." By working in cross-organizational teams, the company has reduced new product development time from 14 to 18 months to 9. "Every insurance company should be doing this," Nason says who also cautions that it requires patience. "It's a long and arduous process. The magnitude of the change we're talking about is so huge that there's often much resistance throughout the organization."

Not surprisingly, virtual teams also are springing up in the very industries driving the momentous changes that are carrying us from one age of civilization to the next.

A SunTeam Success Story

One company betting its future on operating in cyberspace since its 1982 inception is California-based Sun Microsystems. Highly decentralized—it comprises six independent "operating companies"—Sun maintains an extraordinary information infrastructure: 1.5 million e-mail messages flow through the 17,000-member company each day. Some Sun people say they no longer use paper at all. What other companies manage with more people, Sun tries to achieve with better and faster communication systems. CEO Scott McNealy's 1995 corporatewide injunction "to operate on Internet time without compromising quality" set a daunting new standard.

With sales soaring and profits keeping pace with the annual good news, Sun nonetheless launched an initiative that same year to solve some "real nasty problems," as the company's head of research and development, W. R. "Bert" Sutherland, puts it. In a few short months, it created 70 "SunTeams," virtual teams that operate across space, time, and organizations to address a number of critical business issues that the company identified (see Chapter 7).

Launched by Customer Request

Among the problems that the company wanted to solve was how to respond to requests for additional services from large customers. "Motorola, for example, wanted EDI (Electronic Data Interchange) ordering according to their own system requirements," reports Bill Crowley, who co-led one of the SunTeams and who in his "day job" serves as Operations Manager-North America for SunExpress, Sun's aftermarket business unit. "Our challenge was to figure out how to mass customize things that appeared to be highly customized. Could we then promote them as products?"

To solve the problem, Crowley and a few of his colleagues formed the Customer Order Cycle Team. "Phase 1 was to identify the services that customers were requesting and decide which one to work on. We selected Motorola and its EDI ordering system request as the test case. The idea was that they would be able to place orders online for standard things that they use all the time such as toners and cartridges for their printers and have them in two days. They could place their order online, have it checked for availability, and then have it shipped. Minimal human intervention would be required unless there was a stockout or a problem with the order."

The next step was to expand the team to include all the people they needed. Crowley co-led the team with another SunExpress manager, both of whom were based at the business unit's headquarters in Chelmsford, Massachusetts, 3000 miles from Sun's home base in California. "When we first started the team, we hadn't yet selected the program we wanted to implement," Crowley says, "but after we made our choice, we needed to add more people." To cover the company's two sales regions outside North America, they recruited a marketing person from Sun's Japanese operation and one from Sun's European operations in Almere, Holland. In addition, they enlisted finance, information resources, and marketing people from SunExpress headquarters. They also sought the sponsorship of two senior executives, the general manager and the vice president of worldwide operations, both of whom report directly to the president of SunExpress.

"One of the values we had was to involve our customers and suppliers as we needed them," Crowley says. Thus, Motorola's Austin, Texas, operation, which initiated the original customer request, and a supplier, Caterpillar Logistics Systems, based in Peoria, Illinois, which provides transportation and warehouse management worldwide for SunExpress, both became episodic members of the team.

"One of our critical internal relationships is with Sun Microsystems Computer Company (SMCC) (the Sun operating company that designs and produces its products). Every SMCC customer becomes *our* (Sun-Express) customer at some point so we also had one of their sales reps involved. Motorola is a huge account and we wanted to make sure that we were working in conjunction with SMCC sales," Crowley reports.

"We moved people in and out as we needed them, kept senior management up-to-date, and made sure that anyone who was impacted knew what we were doing." The team invited the senior sponsors to meetings when necessary and included them in the regular e-mail distribution list.

E-Mails and Meetings

Remarkably, the team completed its work in seven short months without ever holding a face-to-face meeting for the entire group. Weekly meetings took place via conference calls with people phoning in from their remote locations. "We had offline meetings as required but never had our Japanese member, the Europeans, and everyone else in the room at the same time. We were heavily dependent on e-mail which was our #1 communication tool," Crowley reports. Amazingly, for a company with the technology power of Sun, they never used videoconferencing or any sophisticated online project management software. "We were a small team of 15 rather than 100. Sometimes getting into those highly structured project management systems slows things down."

Agendas were produced prior to each meeting with decision points carefully identified. "Our strategy was that we did the work during the week outside the meeting and then came to the meeting prepared to talk about updates or problems. We very specifically kept our meetings to two hours. That was where the critical role of the team leaders came in, making sure we got through the agenda and did not get stuck."

While pointing out that strict protocols for managing virtual teams such as restricting meeting lengths are important, Crowley also cautions that "it's not necessarily bad to break the rules of the meeting. You can't be too regular about anything. There are no breakthroughs without breaking the rules."

This tolerance for the unexpected is an important feature of working at a distance. Since there is no time-worn body of experience to draw from, virtual team members have to be open to experimentation, often discovering what made them successful in hindsight.

"In retrospect, we realized that we had a formula for success," Crowley says. "Senior management involvement *plus* cross-functional experts *plus* team commitment to the process *plus* stakeholder buy-in equals success."

What Crowley's team did intuitively was to follow the prescription for successful virtual teams:

- They involved the right *people* both from internal organizations and from outside companies.
- They carefully defined their *purpose* and used it as a compass when they started to get off track. "Always keep the end goal in front of the team," Crowley says. "Asking the 'what is the original intent?' question tends to get people back on board in the right way."
- They established excellent communication *links* among the team members, using a mix of media including e-mail, conference calls, and face-to-face meetings to support interactions and relationships.

When the team completed its work, SunExpress had an EDI ordering system and a process in place for responding to new product and service requests from its customers—all in a little over two quarters' time—Internet speed, indeed.

Virtual Team Principles

Work in a world in which the sun never sets is very complex. There are few maps in this new world of work and lots of complaints. People are trying to feel their way, uncertain that they are making the right decisions.

Most of us never received any training for living and working in a fluid, instantaneous, global "village." Thus, we need new models for teams that also incorporate the timeless features of working together.

Three words capture the essence of successful virtual teams:

- People
- Purpose
- Links

People populate small groups and teams of every kind at every level—from the executive suite to the subcommittees of the local school's parent association. *Purpose* holds all groups together, but for teams, the task—the work that expresses the shared goals—is the purpose. *Links* are the channels, interactions, and relationships that weave the living fabric of a team unfolding over time. The greatest difference between in-the-same-place teams and virtual ones lies in the nature and variety of their links.

The People/Purpose/Links model (Figure 1.2) unfolds into nine Virtual Team Principles, which provide a framework for practical, adaptable approaches to the creation and management of virtual teams.

Three Slants on People

- Independent members   *Parts*
- Shared Leadership   *Parts-as-wholes*
- Integrated levels   *Wholes*

Virtual teams comprise *independent members,* people with a modicum of autonomy and self-reliance. Although leadership tends to be

Figure 1.2   Virtual Team Model

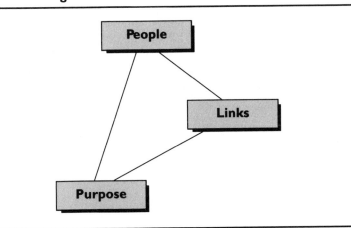

informal, invariably the diversity of technical and management expertise required in cross-boundary work means that most members take a leadership role at some point in the process. In virtual teams, *shared leadership* is the norm. Finally, the team is a human system arising from people parts. It has at least two levels of organization—the level of the members and the level of the group as a whole. Teams also grow out of and are embedded in organizations; they are parts of larger systems. To be successful, virtual teams must *integrate levels* both internally (subgroups and members) and externally (peers and supergroups).

The Point of Purpose

- Cooperative goals   *Do*
- Interdependent tasks   *Doing*
- Concrete results   *Done*

Purpose, which defines why a particular group works together, expresses some minimal level of interdependence among the people involved. Virtual teams are far more dependent upon having a clear purpose than face-to-face teams. Because they operate outside the bounds of traditional organizational life without bureaucratic rules and regulations to guide them, they must rely on their common purpose to stay in tune.

Cooperative goals are what purpose looks like at the beginning of any successful teaming process. This is why so many books about teams begin by focusing on goals. A set of *interdependent tasks*, the signature feature of teams, connects desires at the beginning with outcomes at the end. When a team completes its process, it expresses its purpose as *concrete results*, the measurable outputs of joint effort. These three elements—cooperative goals, interdependent tasks, and concrete results—enable virtual teams to stay focused and be productive.

The Web of Links

- Multiple media   *Channels*
- Boundary-crossing interactions   *Communicating*
- Trusting relationships   *Patterns*

What gives *virtual* teams such distinction is their links. Relatively suddenly, multiple, constantly enhanced modes of communication are widely available, providing access to vast amounts of information and unprecedented possibilities for interaction. We chose the term *links* for this defining feature of virtual teams because it bridges three key aspects of communication.

First people need the actual physical connections—wires, phones, computers, and the like—that provide the potential for communication and are the prerequisite for interaction. *Multiple media* are moving virtual teams from the extraordinary to the ordinary as the technology wave of Information Age change reaches the mainstream.

Connections make *boundary-crossing interactions* possible. The back-and-forth communication between people—the activities and behaviors—constitute the actual process of work. It is here—at the boundaries of interaction—that virtual teams are truly different.

In virtual teams, people's interactions across boundaries require behaviors that are fundamentally new.

Through interactions near and far, people develop *trusting relationships*, the invisible bonds (and baffles) of life. People's patterns of behavior mark the outlines of their relationships that persist and feed back into subsequent interactions. As important as positive relationships and high trust are in all teams, they are even more important in virtual ones. The lack of daily face-to-face time, offering opportunities to quickly clear things up, can heighten misunderstandings. For many distributed teams, trust has to substitute for hierarchical and bureaucratic controls. Virtual teams with high trust offer this valuable social asset back to their sponsoring organizations for use in future opportunities to cooperate (see Chapter 9).

It is far better to cross boundaries than to smash them.

Cyber Teams

With electronic technology, virtual teams can work across time and space in ways that provoke the formation of entirely new ways of working and organizing. The word *cyber* is telling: it means steersman in Greek, putting you in the driver's seat. To travel across distances faster than the speed a person can walk requires transport—ships that sail across oceans to new worlds, horses that ride over mountain ranges to new frontiers, electronic media that open up cyberspace.

The World Wide Web Inside and Outside

In an area of fast paced technology change, such as communication, it is dangerous to make predictions. We do, however, put a stake in the ground for the awesome impact of the Internet and intranets on the future of virtual teams.

The particle physicists at CERN (Conseil Européen pour la Recherche Nucleaire)[9] in Geneva, Switzerland, came up with a powerful new way to interconnect their global research community using the Internet in 1989. When they did, they could not have predicted what would happen with their technology. In a few short years, the World Wide Web has become the communication phenomenon of the decade. Suddenly anyone with Internet access can explore millions of postings around the world on nearly every conceivable topic gliding from continent to continent in click-of-a-mouse time.

Ted Nelson and Doug Engelbart were among the earliest seers to envision the possibilities of *hypertext* and the resulting global web of networked knowledge. The word *hypertext* is rather self-descriptive: It is text that behaves as if it is hyperactive. Anything written in hypertext (or any graphical element) can become what amounts to a computer button of its own, the simple but profound linking capability. Click your mouse and off you go to that destination. This means that everything on the Web can be interrelated—linked. Thus, witness our ability to wonder what country the Kalahari Desert is in and discover the answer—Botswana—in less than 30 seconds, thanks to the World Wide Web.

The technologies that support hypertext make the World Wide Web possible:[10] The basic computer language used to design Web pages is called HyperText Mark-up Language (HTML); the communication link that connects Web sites uses a standard called HyperText Transfer Protocol (HTTP).

Companies, libraries, universities, government agencies, hobbyists, nonprofits, political campaigns, social activists, and just plain folks all have jumped on to the World Wide Web. Millions of home pages are joined by many thousands more every day.

For virtual teams, the use of private World Wide Web sites is a singular blessing.

> *For the first time, teams can virtually collocate all the information they need to work together and put it all in context.*

That does not mean that all of the team's information is on the Web site, but it does mean that the Web site can point you to wherever you need to go. Pointers are embedded in text, outlines, graphics, maps, and other media. Virtual reality for virtual teams is going digital and becoming conceptual.

When Microsoft bought Vermeer, a small software company whose product, Front Page, is a Web page design tool, the two companies needed to do a lot of work to integrate the Boston-based firm into the Redmond, Washington, giant. Instead of using the familiar frustrating process of endless meetings, phone tag, and lost memos, Richard Dale, then Vermeer's vice president of Operations, used a different solution. "The work of moving Vermeer was a pretty mundane sort of thing requiring a bunch of fairly boring tasks," he explains. Dale's observation echoes the experience of many people involved with such transitions—closing offices, arranging the move, resolving personnel issues.

So Dale decided to liven things up. He used his company's product to design "a little Web page which had all the numbers we needed. Anything we needed to remember we put on our internal Web site."

"Intranet" is the term for a private Web site that is internal to an organization or one of the networks that make up the truly vast Internet. Dale's company already had a rather substantial intranet running. It contained everything from "project status to people's names and addresses to forms for ordering office furniture to how to send a FedEx, to what was on the lunch menu at the local take-out place."

"I was in charge of operations for the company," Dale says, "and my philosophy was that if more than one person asked me a question, I put it on the Web site." The fancy name for Vermeer's intranet? "The Internal Web Page." Soon their Web page became Vermeer's institutional memory. They also had an external page for customers.

The Vermeer home page (as the "front door" to a Web site is known) had a norm attached to it that underscores the subtle organizational difference that virtual teams reflect. Everyone in the group had privileges to "author," that is write, material to the page. "That wasn't controlled, but we were only 35 people and we knew that if people started to mess it up, we'd control it," Dale explains. "The day we were bought someone put up a picture of Bill Gates and set a link to the Microsoft home page. The due date for one of the engineer's baby became a sweepstakes."

There was even a Coffee Manifesto, which included instructions on how to use the coffee machine on the site. (As Dale says, "You know software engineers are very particular about their coffee.")

The family feel to all this speaks to the sense of community that the technology makes possible, that elusive quality called virtual that makes work-at-a-distance possible, allowing previously separated people to be pulled into a team. While older qualities of virtual teams find new expressions in cyberspace, true novelties also are appearing.

Information Is Money

"Rocks Bank," a pseudonym for one of the world's largest custodial banks, handles $500 billion annually, about a quarter of the entire budget of the U.S. government. Until recently, one huge player—State Street Bank—and two smaller ones—Chase Manhattan and Bank of New York—dominated custodial banking. When the two smaller players, each a giant in its own right, grew separately through mergers and acquisitions, the industry became explosively competitive.

Suddenly custodial services—all the backroom functions that enable money to change hands in mutual funds, pension funds, and private transactions so they can finally rest someplace (in custody)—became a commodity. To distinguish themselves to their clients, custodial banks have become information providers. The boundaries between financial services and information technology services have faded in an electronic instant.

Rocks Bank's clientele is diverse, including mutual funds and state and corporate pension plans, one of which is an investment company that is both its client and its competitor. "We compete fiercely with them for 401(k) management but we're also their custodian," says Ian Stevens, Rocks Bank's vice president of New Technology. "We hold about $50 billion of their money because they're not a bank. It's all so interwoven and incestuous." Even clients and competitor lines have blurred for this institution that is growing at astronomical rates—25 percent annually year after year.

If you think the most exciting software development is coming out of the brand name houses or even the brilliant boutique shops, reconsider.

Places like Rocks Bank also are building the electronic superstructures for virtual teams.

Stevens is the architect behind Rocks Bank's Work Stream Technology project, a harbinger of technology support systems for virtual teams. When complete, the system will make each trade "a self-aware work object." The trade will move around the world among virtual team members according to what it knows needs to happen to it next on its own clock. ORBs (technically, "object request brokers") at each site coordinate and distribute work in a global 24-hour-per-day office that is always open for business. Object technology like this, the next generation beyond relational databases, eventually will make virtual teams completely self-sufficient. When Rocks Bank's home office in Sydney, Australia, closes at 5 PM, its ORB will automatically move work to its New York branch where that part of the virtual team is just waking up. The individual work object will carry its own set of instructions complete with the attached interactions of everyone who has worked on it.

Every time a trade is made on any of the assets that Rocks Bank holds for its clients, the bank needs to know about it. Simple enough, but 50 percent of the transactions that the bank receives in its Medallion Trust department each day come in by fax—many of them handwritten! The fax problem was only the first obstacle that Stevens' team had to overcome: Medallion Trust, just a small part of the bank, receives 2000 faxes a day. Each fax has its own demanding timetable and elaborate set of actions that its existence initiates.

"In this one department, we have 50 people and half a dozen fax machines that run nonstop with a very complex method of logging and tracking," Stevens explains. "Faxes are frequently either missing information or they are illegible. 'Is that $5m or $50m?'"

So his team's work is to build an environment that will allow each transaction to become its own self-collecting history. By working around the clock without disruption, passing work from time zone to time zone, the bank can increase its volume and accelerate its service—quite a competitive advantage. The team will be truly virtual—their hand-offs reaching around the globe following the path of the rising sun.

Feeding the Virtual Team Cycle

Virtual teams are not just a nifty way to organize and make use of cutting-edge technology. Whether consciously or not, many companies like Rocks Bank are betting their future on virtual teams as their strategic differentiator. By employing virtual teams, they can do things that are impossible within the prevailing model of side-by-side, 9-to-5 work.

Virtual teams are a strategy for success.

If they cannot accomplish their goals within their own four walls, virtual teaming companies climb over them and partner with someone or several someones with whom they can make it happen. If their competition suddenly overpowers them, 21st-century organizations see virtual teams as the way to become smarter and more flexible, adaptive and more competitive.

The way is not easy. Virtual teams are microcosms of the organizations and environments that spawn them. Today's teams are complex and reflect all the stresses and strains induced by the extraordinary shift in human civilization now underway. As the Industrial Age recedes more swiftly and the peak of the Information Age still looms far ahead, we and our groups are betwixt and between. We are born into ages past, yet navigate ahead to an uncertain future.

The Virtuous Loop

Many teams now are physically distributed. Long-standing management molds that funneled information up and sent orders down are cracking. More information is becoming more omnipresent to more people. Competitive pressures to constantly improve cost and quality are driving the redesign of work processes. All the while, information seeks its natural path, flowing with its own simple process physics, horizontally linking people across boundaries through and among organizations.

There is a *virtuous*[11] *feedback loop* building in the development of virtual teams that promises an exponential rise in this form of organization. Virtual teams are not a fad. They are the future.

The virtuous loop begins with yesterday's assumption that people must collocate to work together. "Shoulder to shoulder" the traditional team works together. Shoulder to shoulder traditional teams hand off their work to the next team in chains of larger processes, the bucket brigade view of working groups. Organizational building blocks of closely spaced bodies stacked in command-and-control pyramids. This is the idealized machine organization of the Industrial Age. Thus:

- **Given:** Traditional work group design creates stable spatial and organizational boundaries based on locating people with interdependent tasks next to one another.
- **Change:** Today, however, technology, speed, globalization, and complexity are rearranging this root premise of work design.
- **Impact:** As a consequence, people working on interdependent tasks are no longer necessarily proximate in space and time, nor need they be in the same organization. Two things happen: distance and time become problems to solve and organizational issues develop within rigid hierarchy-bureaucracies. To deal with distance, people usually turn to a mix of face-to-face meetings and electronic communication technologies to replace key elements of collocation. To deal with the demands of cross-boundary work, organizations create virtual teams as needed.
- **Novelty:** Electronic, particularly digital, media that people typically use to compensate for distance, eventually go beyond their replacement application. In time, they offer entirely new ways for people to communicate interactively.
- **Adapation:** This in turn leads to new networked forms of organizations—which are virtual teams at the small group level. Meanwhile, new energy is pumped into the system as increasingly more work is created around digital products and services.
- **Result:** As the technology and organizational support structures for virtual work improve, more work is designed to take

advantage of new network technology and management. This only fuels the trend toward virtual teams, making it easier for them to work interdependently on tasks that cross boundaries, feeding the loop. So, more and more virtual teams are in our future.

Looking Forward and Back

A good virtual team is, at its heart, a good team. Since many virtual teams do meet periodically or a few times or at least once, they also find themselves in the conventional face-to-face setting. As Bernie DeKoven, a pioneer in using technology to support virtual teams,[12] says, "When I think of virtual teams in the best light, I think of teams of people who are as comfortable with each other as they are with a wide variety of communication and computing technology. When they meet 'virtually,' they take advantage of all their technical know-how to continue their work; and, when they meet face-to-face, they use the same technology to develop, organize, and refine their understanding. They have an emotional bandwidth that is as broad as their communication bandwidth, so that no matter how or where they meet they relate to each other with humor, understanding, and respect."

For virtual teams to be complete, they must include what is timeless and enduring in human groups. They also must include the features that are really new in the turbulent years at the turn of the millennium. The challenge of our time is to invent and improve virtual teams and networks while retaining benefits of earlier organizational forms.

CHAPTER 2

TEAMING FROM THE BEGINNING

How Groups Became Virtual

"Long Distance Operator,
This is Memphis, Tennessee."

Chuck Berry's famous 1958 hit could be the theme song for an unusual scenario. Cameron Harker, an Australian chemical industry salesperson, has an urgent question about a trial he is running for a client at a Brisbane paper mill. He knows that someone somewhere else in the world probably has the answer. Fortunate to be an employee of the specialty chemical company Buckman Laboratories based in Memphis, Tennessee, he simply sits down at his keyboard, clicks on his modem, and taps out a message. Within a few hours time, his problem is solved. Responses come back from people around the world—Canada, South Africa, Sweden, and the United States. They also come from many levels in the company—a general manager, a national sales manager, a product development manager, and people in research and development.

Our Company Never Closes

In the new boundary-crossing world that Buckman Labs (as it is known) inhabits, Cameron Harker does not actually dial into company headquarters in Memphis. There flags from each of the countries in which it has subsidiaries fly at the main entrance. Instead, he makes a local phone call through which he accesses one of Buckman's 62 discussion areas on a global electronic communications network. By dialing in and posting his request, others among his 1200 company colleagues will see his message and respond immediately—even though they live in 19 countries on 5 continents with customers in 80 countries.

The city famous for Elvis, Beale Street, and the Blues is also home to a company now celebrated in virtual team circles. Buckman Labs has been way ahead of the pack for a long time with its problem-solving teams that circle the globe. After more than a decade of pioneering its online knowledge network, Buckman has answered a lot of questions that other companies are just beginning to ask. And, the city of Memphis, known as "the distribution center of the United States," itself is no laggard. Because of its geographic centrality, many companies, including Federal Express and Northwest Airlines, use it as a hub. A test bed for many of BellSouth's new telecommunications products, the city has had high-speed data lines (called ISDN) available since the early 1990s.

Increasing the Span of Communication

Bob Buckman is the visionary behind Buckman Labs' global "knowledge transfer" system, K'Netix. In 1996, Buckman stepped down to vice chairman after 17 years at the helm of the privately held company that his father started in 1945. A $270 million company, Buckman Labs provides specialty chemicals to the pulp and paper, water, and leather industries. The company's products keep swimming pools free of algae, clear effluent streams of heavy metals, and allow contamination-free manufacture of paper products.

The online network that allows everyone in the company to talk to one another from anywhere at any time is not just supporting technology for the company. It is the logical manifestation of Buckman's deeply held beliefs about people, business, and the nature of competition. Buckman often begins his speeches with a telling quote from former Scandinavian Airways chair Jan Carlson:

> *An individual without information cannot take respon-*
> *sibility; an individual who is given information cannot*
> *help but take responsibility.*

For people to be effective, Buckman says they have to have information with which they can increase their "span of communication" and thus their "span of influence."[1] These key tenets that lie at the heart of the K'Netix design allow and encourage everyone to participate. Although he is a chemist and a statistician, Buckman sounds at times like a communications theorist. In this case, good theory appears to make for good business: the number of sales from new products has grown dramatically since the early 1990s when the online information system began to burgeon.

"The speed at which you can communicate defines how quickly you can make money," Buckman says. "If I can respond to a customer in six hours anywhere in the world at any time, that's a competitive advantage. As the speed of communication increases, customer response time moves toward instantaneity. That redefines competition. Any entrepreneur in the world will understand that."

To unleash the power of the individual, Buckman says to "radically change their span of communication and I mean radically. Anyone should be able to talk to anyone else inside and outside the organization. We want to close the gap with the customer. How do we increase our cash flow with the customer? By increasing our power on the front line. But that can only happen if the individual has good span of communication."

Buckman's goal is to have 80 percent of the company "effectively engaged on the front line," that is, directly connected with customer needs. "If you're not doing something useful for a customer, why are you here?"

Currently, he figures that 50 percent of the company is on the front line, an increase from 16 percent in 1979.

The impetus for K'Netix was the desire to share best practices for solving customer problems. "But we couldn't run Ph.Ds around the world fast enough at the speed that we needed," Buckman recalls. When they started the network in 1985, a portable PC weighed 17 pounds and e-mail was in its infancy. "That's where it all started—with e-mail, trying to communicate better, pure and simple."

Geographic separation was not Buckman's only problem. Even when people are located in the same place, Buckman figures that people are out of their offices on average 86 percent of the time. "Why do organizations spend huge sums of money on systems that only function 14 percent of the time?" Buckman asks. Most salespeople spend practically no time, if any, in the office. Likewise, executives and most managers are rarely at their desks from 9 to 5 every day.

The desire to share best practices *anywhere all the time* turned into an online discussion system based on some very simple principles. It should:

- Reduce the number of transmissions of knowledge between individuals to one.
- Give everyone access to retrieving from and contributing to an easy-to-use, fully searchable, automatically updated company knowledge base.
- Be available all the time anywhere in the world since "our company never closes."
- Communicate in whatever language is best for the individual user.

Maximizing Participation

By today's stupefying technology standards, the Buckman global knowledge network is pretty elementary. "We use the same e-mail system that 4.5 million other people do," says Victor Baillargeon, the Ph.D. chemist who was until 1996 vice president of Knowledge Transfer at Buckman. After five years using a cumbersome IBM network that required different codes for different countries, they moved in 1992 to CompuServe, the Columbus, Ohio-based network. Everyone at Buckman can easily dial into a local service or directly into a CompuServe node.

CompuServe "forums" have been around since the mid-1980s. The idea behind the forums is very simple. Each one is on a different topic and anyone with forum access privileges can post messages to it. Because it captures everything electronically, the system maintains its own ongoing history of the discussion. Buckman has numerous forums on topics germane to its business such as the TechForum that has 24 discussion groups (including BuLab News, the online company newsletter) that are open to everyone in the company. The Purchasing Section is where all Buckman purchasing agents around the world communicate. The forums also house proprietary discussion groups. There new products, corporate strategy, and finance are under discussion by small teams with direct responsibility for these areas.

"We chose CompuServe because of ease of use," explains Alison Tucker, who managed the forums for the first several years and now heads Buckman's Internet initiatives. While other companies are out hiring programmers to develop complex information systems, Buckman chose instead to buy off the shelf. "The only reason companies create their own systems today is ego!" Bob Buckman says emphatically.

The Buckman system works because the barriers to participation are low. Everyone has access to PCs and laptops with modems. When people travel, they can take an "electronic first aid" kit with them, equipped with whatever they will need for the country they are visiting including adapters and cables.

How do you get 1200 people around the world comfortable with logging in every day to solicit and contribute advice to people they never (and may never) see?

It took a lot of time, training, attention, and senior-level commitment. At the beginning, Tucker spent "endless amounts of time online. Sometimes 12 hours a day," she recalls. "I was learning to manage all these crazy discussions. When we started out, half the things going on were not business related—people talking about their kids and their dogs. We've always been this global company, but people didn't have a chance to talk until this happened. We'd have [real-time online] chat sessions at 5:30 or 6:00 PM Memphis time with people saying, 'I have to go answer the phone or door.' People in Japan were telling jokes to people in Brazil."

To encourage companywide participation, Buckman himself took to the world circuit. He gave speeches on the role of knowledge transfer and his belief that the company's intelligence lies "between the ears of the people, not in some database." From the beginning, he has been an active daily participant. "It has to have unequivocal leadership at the top," he says.

For the first six months, they ran weekly reports to see who was participating and who was not. "Bob sent very civil, gentle messages asking people why they weren't online and whether they needed any help," Baillargeon recalls. "But the unspoken message was very loud and clear. If you wanted to keep your name off the report that was run on Fridays, then you logged in by Thursday afternoon. We haven't needed to run a report since the first six months."

Leveling Out

The initial getting-to-know-one-another frenzy lasted for about three months and then things began to settle down. Today, the system still has a Break Room that has no monitoring and no maintenance. There the Super Bowl, the World Series, and World Cup rugby (which "sometimes gets pretty heated among the Irish, Australians, and South Africans," according to Baillargeon) are the burning topics. Buckman built in permission to use any feature of CompuServe's system from the beginning. The company has encouraged nonwork-related online exploration as an important way to learn the electronic environment.

Over time, the conversations have become self-policing. Baillargeon reports few incidents of "flaming," where people vent, argue, or attack one another online. Participants have developed a style of communicating their requests that is to-the-point and informative. Skilled experts on the topic moderate all business forums. They monitor the discussions, provide help where needed, and archive material that becomes part of the company's knowledge library.

"Our use was very high at the beginning while everyone was learning and now it's leveled out," Tucker says. "The key thing is to be patient and advertise, advertise, advertise to your people. We didn't have anybody to learn from. And we still don't, but we're learning from each other."

K'Netix is the overarching term for all the electronic environments of the company, not just the online forums, which are considerably more than online question-and-answer sessions. "These discussions aren't just a few snap answers," Baillargeon says. "It may take days or even weeks until subjects are complete. A lot of the time it's a full-blown discussion, not just here's a question, here's an answer."

A Climate of Trust

Buckman people agree that such a system would be impossible without a "climate of respect and trust that has to be pervasive," in Baillargeon's words. "If I'm asking a question from one part of the world and I get a reply from someone I don't know in another part of the world, I have to trust that they're giving me their best effort and their best knowledge. Part of my incentive to participate is that today I may be bothered by having to reply but tomorrow I don't know what I'm going to need to ask for."

The "climate of trust" at Buckman is represented in a "Code of Ethics" that new employees receive in laminated pocket-size form and which hangs framed in every office.

> *Because we are separated—by many miles, by diversity of cultures and languages—we at Buckman need a clear understanding of the basic principles by which we will operate our company.*

So begins the firm's social compact born out of a very practical problem that arose in 1983 when the company's global presence started to become significant. "Folks working in other countries started to ask what they should tell their people to do about bribes," Buckman recalls. "Everyone sent in ideas and Steve Buckman [Bob's cousin and, since 1996, CEO and chairman of the company] pulled them together and it's never changed since. It became the glue."

The Future of K'Netix

After 10 years of learning, Buckman Labs continues to push out the electronic border. Finance, human relations, and a host of other corporate services have moved completely online. "If the SEC (Securities and Exchange Commission) feels secure using CompuServe, then we should," says Buckman when asked about security issues. A marketing information data analysis system is under development as is a Customer Information Center, which will be the repository for all customer information. Forums are being developed that include customers, and the entire database is being migrated to icon-based images. World Wide Web pages are materializing to deal with a wide variety of topics and a user-friendly interface is being built to the groupware product Lotus Notes.

Perhaps most interesting is the Distance Learning project that Buckman himself chairs. The idea is to bring knowledge to the learner rather than bring the learner to the knowledge. Since the Buckman learners speak nearly a dozen languages, the frontier issue of instantaneous translation is the current problem to solve. Because the company's business of producing specialty chemicals is so highly "specialized," construction of the dictionaries for translation is its own headache.

Bob Buckman is not stopping at simply providing distance learning in the native speaker's language. His hope is that eventually K'Netix will be entirely language-sensitive. When someone posts a note in English, people in Japan will be able to read it and respond in Japanese which in turn people can read and respond to in Portuguese, Swedish, Dutch, French, Spanish, and Italian as well as English—and whatever other languages employees speak.

In the end, Buckman believes that the system's success rests on people not tools. "It's 90 percent culture change and 10 percent technology," he says. "You cannot drive this change through technology and technology budgets. It's people who bring about the change. These systems and methodologies are now self-generating and don't need that long, hard push to indoctrinate into the culture. It's a journey not an end and we are still struggling."

Global teams that never meet are the latest in a long line of innovations in small groups. We have been working on this form of organization for a long time.

Four Ages of Small Groups

The magnitude of the change now gripping us all—in speed, complexity, and globalization—is captured by the view that humanity is undergoing a major evolutionary transition from an industrial to an information-based economy and society. Alvin and Heidi Toffler use "three waves of change"[2] to capture the essence of the big view of human civilization:

- The Agricultural Age wave began 10 to 12,000 years ago, and marked a dramatic shift from the Nomadic Era. Farming and herding eventually replaced hunting and gathering. Populations grew larger, cities and towns developed, and family size increased as people settled down.
- The Industrial Age, running roughly from the 18th through the mid-20th century, saw factories replace farms as the economic engine. Populations have exploded and urbanized, while families have grown smaller. This age represents today's tradition, the old from which the new seeks to emerge.
- The Information Age is growing out of the third wave of change, beginning in the mid-20th century. We are now riding the turbulence of transition. The world's economies are becoming information-based, electronically connected, and globally interdependent. Population is still rising and families are still small but diversified.

The Tofflers' three waves of change divide all of human history into four great ages. To span all our contemporary organizational capabilities, we need to include the first human era.

- The Nomadic Age, beginning indistinctly between two and three million years ago, was when our ancestors acquired the ability to speak, make tools, and configure social organizations. Populations were sparse and families were relatively small.

Four Varieties of Organization

Each of the great ages also has initiated a new social configuration.[3] Nomads roamed around in *small groups.* The great agricultural organizations were the first *hierarchies.* The rise of industrialism brought the large-scale use of *bureaucracies.* The Information Age has its emblematic organization as well: boundary-spanning *networks* (Figure 2.1).

Over the ages, we have accumulated organizational knowledge. When hierarchy came along, people did not stop meeting in small groups. When bureaucracy evolved, hierarchs did not throw down their scepters and call it a day. Indeed industrial bureaucracies depend upon the ranks and levels that agrarian hierarchies invented. While developing its own signature characteristics, each age also incorporates essential organizational features of the ones before it.

Networks, the emerging organization of the Information Age, incorporate aspects of its predecessors: the levels of hierarchies, the specialties of bureaucracy, and the purposes of small groups.

Old forms do not, however, persist unchanged. With each new age, new versions of old forms supplement the human organizational repertoire.

Figure 2.1   Ages of Organization

Nomadic	Agricultural	Industrial	Information	Age
Small Groups	Hierarchy	Bureaucracy	Networks	Organization

*The original autonomous small group—the family—
survives today, still central to society yet different in
each era.*

New forms of hierarchy (for example, shared leadership at the top)
and bureaucracy (for example, decentralized) are appearing within networked organizations.[4] There also is a new variant on small groups—
virtual teams (Figure 2.2).

Groups of Nomads

The mobile family that foraged to survive was the basic social unit of the
Nomadic Era. Relatively small in size, these families were partly self-
sufficient and partly interdependent with other families. Together, they
periodically set up camp in larger groups. Once in the camps, task-
oriented groups naturally took shape. Hunters, gatherers, and traders
joined forces according to circumstance and need.

Figure 2.2   Four Ages of Small Groups

	Nomadic	Agricultural	Industrial	Information
Families	Mobile family	Extended family	Nuclear family	Diverse family
Task Teams	Gatherers Hunters Traders	Farmers Herders Artisans	Positions Specialties Professionals	**VIRTUAL TEAMS**
Social Groups	Health Leisure Friendship	Castes Classes Religious	Associations Special interest Clubs	Electronic groups Virtual communities
Decision Groups	House heads Camp councils	Rulers, elites Military units Owners	Legal Representative Committees	Direct participation Virtual government

> *The first teams were the task-oriented camp groups of the Nomadic Era.*

Camps also stimulated other relationships outside the family. These nonkin affinity networks, as the anthropologists call them, enabled people to have friends, share information, offer healing, encourage hobbies, enjoy leisure and recreation, and participate in contests. Without cooperation across kin lines, humanity never would have gone beyond subsistence. These "virtual" kinships, what anthropologists call fictive kinships, were critical to human progress.

The Agriculture of Small Groups

In the Agricultural Era, families grew to be larger and more extended. Farmers and herders, the new task-oriented economic units, eventually crowded out hunters and gatherers (although not completely, hunter-gathering societies still survive on several continents today). Skilled tool-makers evolved into artisans. With them came the masters of the trade with their own small shops and apprentices. Society stratified into castes and classes. Religious groups emerged as a common spiritual life integrated larger communities inhabiting bigger societies.

The great organizational innovation of this era was the rise of ruling elites and military units. To protect land, agrarian settlements marshaled military hierarchies that could coerce people. Along with organized violence, hierarchy brought along a positive development: The clear efficient authority structure of ranks, small group units combined into larger units. This innovation—which uses the cross-systems principle of levels—was a great leap forward in the human capacity to organize large numbers of people.

At the same time, the first cornerstone of capitalism was put in place. Military and religious leaders became owners of land previously held by the groups who lived on it. In economic terms, ownership is the ultimate source of coercive authority—the right to sell, to hire, and to fire.

This feature is the bedrock of hierarchical rights and responsibilities in companies and it affects teams today. Because they are by their nature hierarchical decision makers, executives must confront a schizophrenic problem when they try to team with one another. They have a dual nature—hierarchs who are guardians of supreme power and team members who are partners in power. This makes it particularly difficult for executive groups to become effective task-oriented teams.

Organization as Machine

In the Industrial Age of the past few hundreds of years, the typical family size shrank again and became more nuclear. While remaining key to the social domain, the family abruptly ceased being the basic economic unit of society.

Instead, segmented, specialized work ruled. Task-oriented bureaucratic units became the basis for economic gain. Rules bound replicable operating units. The units in turn aggregated into larger mechanical processes that produced predictable results. Society viewed small groups of all sorts as interchangeable, replaceable parts of the machine organization.

Formal representation under law, where small groups stand for larger communities of people, is the great social invention of bureaucracy. New organizations defined by constitutions, laws, policies, and procedures created numerous bureaucratic small group structures—from Supreme Courts to city councils.

Small Groups in the Information Era

As the millennium draws to a close, increasingly diverse styles of families are proliferating. The nature and role of the family are hot topics. Families are again becoming a significant economic unit, not just as consumers but as joint "businesses" with two or more income streams.

At work, distributed, decentralized, flexible organizations are replacing many collocated groups. The technological capacity to share information and the staggering increase in the ability to communicate provide fertile soil for the growth of *virtual teams*, the new boundary spanning, task-oriented working groups born of the Information Age.

Online affinity groups attest to the emergence of virtual social groups, virtual communities.[5] Still hard to see at this point are the new governance forms. As the ultimate protector of "the way things are," and bound by laws, rules, and regulations too numerous to count, government is likely to lag behind in organizational change. Yet even in government, inventive, more flexible structures are proliferating.[6]

The Generic Small Group and Team

Over thousands of years, human life has spawned many kinds of small groups. To see what is special about teams we need to understand what is common to all small groups. Then we can distinguish virtual teams from other kinds of teams.

People tend to have their own definitions of teams and how they differ from groups, but researchers have forged considerable agreement on the matter.

Foundations of Small Groups

As recently as the mid-1980s, the standard model[7] of small groups required a hunt through research in anthropology, sociology, organizational psychology, and management. At that time, there was a building consensus that a small group was a coherent system that one could study independently at the crossroads of several disciplines, but it was not a well-developed field.

A decade later, a search of the literature on groups and teams turns up a coherent field of research. Today, a very clear model of small group characteristics stands with considerable consensus behind it. Indeed, the general definition established in the mid-1980s has enjoyed a decade of testing and exploration. A 1996 summary of current research quotes a respected researcher's 1984 review of the literature with approval: [8]

Virtually all definitions of the term small group *include three attributes: two or more individuals, interaction among group members, and interdependence among them in some way.*

This leads to a very short definition of a small group:

Individuals interacting interdependently.

People become a group by virtue of doing things of mutual benefit together. A small group is *not* a random collection of people, like a crowd crossing a street or passengers on a plane. *Groups* of people have something more, an interrelatedness and a common motivation that adds up to more than just a bunch of individuals.

A collection of people becomes a group when the whole is greater than the sum of the parts.[9]

The standard model embeds a small group in a larger context. It is very rare that a new small group arises out of nowhere. Usually, small groups arise from pre-existing groups—the special accounts group that grows out of a finance organization, the book club that is adjunctive to the church, the pick-up basketball team that grows out of the playground-building project spawned by the PTA. In business, small groups are invariably part of a larger organization or set of organizations.

Individual members of the group define its boundaries. Whatever it is that enables people to say they are "in" the group while others are "out" of the group identifies the boundary. When people are on an e-mail distribution list, they establish themselves as members of that virtual group. If you are not on the list, you are not a member. Membership recognized by insiders and outsiders alike gives a group its basic boundary.

The second element of small groups is *interaction*—the multiple links among members. Communication, the foundation for human interaction and relationships, is inherently a shared activity. Language, first invented in the earliest forager camps, continues to be concocted in new groups today. Acronyms, stock phrases, and in-jokes are all linguistic indicators of group cohesion.

For millennia, small group communications meant that people talked to one another face-to-face, using the medium of sound waves traveling in air. Even today people assume that most small group communication is face-to-face. That is changing.

> *We are experiencing the most dramatic change in the nature of the small group since humans acquired the capacity to talk to one another.*

Researchers prefer the term *interdependence* to describe what most management and popular organizational writers refer to as an essential attribute with terms like "unifying purpose" or "shared goals." The words "individuals interacting" are not sufficient to define a small group. There must be interdependence—joint purpose and shared motivation—to incorporate individuals into a group whole.

Teams Tackle Tasks

So what are teams? In the standard model, the step from small groups to teams is short and simple. Both the scientific literature and the popular press express the distinction in the same clear way:

> *Teams exist for some task-oriented purpose.*[10]

The orientation to task is what distinguishes teams from other types of small groups such as family households, social groups, and governing bodies. While all small groups carry out tasks to some degree (as well as make decisions and support social interactions), task *is* the focus for teams. All other aspects are ancillary. [11]

While having a purpose is fundamental to all small groups, teams are specifically and deliberately results-oriented. Tasks are the work, the common process that is the means to the results, the jointly held end. In setting goals, teams project desired results and agree upon tasks to carry them to their objectives.

In addition to the membership boundary found in all small groups, tasks create a team boundary. The nature of the goals and the work required to carry them out drive the need for certain members and skills to be part of the group. Conversely, different members shape and reshape the purpose and tasks of the group. Indeed, the goals and tasks often exist before the team identifies its members. The feedback loop between task definition and appropriate membership becomes a core defining process during a team's early development.

Purpose in all its forms—interdependence, vision, mission, strategies, goals, results, and, especially, tasks—lies at the center of understanding teams. Purpose also is notoriously difficult to grasp and make predictably practical. Much of the best thinking around teams has gone into books that elaborate various aspects of purpose, in many cases making that thinking applicable through tips, tools, and processes.

While the task focus distinguishes teams from small groups, the essential distinction between teams and *virtual* teams comes in the boundary-crossing nature of their interactions. The day-in-and-day-out reality of communicating, interacting, and forming relationships across space, time, and organizations makes teams virtual.

Virtual Teams Cross Boundaries

Buckman Labs is a sea of virtual teams that constantly form and dissolve. To solve a customer problem, a global virtual team comes together without anyone chartering it and it includes anyone in the company who chooses to participate on a particular topic. When the discussion is over, the virtual team disbands.

Buckman's teams are quite different from traditional task groups comprising people from the same organization functioning in the same place at the same time. This is a description of both the conventional 9 to 5 office and the assembly line 7 to 3 shift that structures the industrial model of work.

Since we are in new territory here with a new type of team, we need some coordinates to explore the terrain (Figure 2.3). One point of reference is the familiar one of space and time. Less obvious but equally important is the dimension of organization.

Space and time are treated here as a single interrelated idea (spacetime) that is appropriate to the concepts of post-Newtonian physics. Distance in space—even a short distance—takes time to cross. At greater distance across time zones, day turns into night and impedes people's ability to interact simultaneously—even with media that travel at the speed of light.

Spacetime and organizational boundaries mix and match into four kinds of teams, one conventional and three virtual, all of which Buckman Labs uses.

In traditional Industrial Age collocated teams, people work side-by-side (the same space and time) on interdependent tasks for the same organization. Several decades ago, even before the new communications technologies were widely available, one form of virtual team began to

Figure 2.3   Varieties of Teams

| Spacetime | Organization | |
	Same	*Different*
Same	Collocated	Collocated Cross-Organizational
Different	Distributed	Distributed Cross-Organizational

appear, the cross-organizational team. Today, cross-organizational teams of all kinds are common.

Collocated Cross-Organizational Teams

Collocated cross-organizational teams comprise people from different organizations who work together in the same place. At Buckman Labs, an informal group of people who number "between 4 and 15," according to Buckman Labs' Alison Tucker, all work in Memphis but for different Buckman organizations. Though their task is to brand the company's products, only some members of the team are from marketing.

Perhaps the most familiar type of virtual team is the classic cross-functional group of experts and stakeholders who come together to solve problems or seize opportunities that require cooperation across organizational boundaries. A good example is the Shell Offshore project team that developed the process for designing, building, and running drilling operations and pipelines a mile undersea. The team included a geophysicist, paleontologist, drilling supervisor, production superintendent, construction engineer, human resources training manager, and organization development consultants. They left their regular jobs and gathered in New Orleans to devote full-time to designing the social and technology systems to support the new operation.

Most of the great management movements of the last two decades have stressed the formation of cross-functional teams. The quality movement and later re-engineering have stimulated the sprouting of teams everywhere, many of them cross-functional. Concurrent engineering and CALS[12] a decade earlier encouraged the formation of new product design and development teams that reached across the life cycle, from marketing and engineering to sales and support, which Boeing used in its 777 design teams.

Although failures are legion and frequent, cross-functional teams (which have permeated every industry, sector, and level of organization) have been an astonishing management success story. The innovation has now spread well beyond internal boundaries. Cross-organizational teams include suppliers, customers, and even competitors in alliances of all kinds that cross corporate borders.

As distances expand across space and organizations, the options to collocate teams shrink even as the ability to work at-a-distance grows. Relocation itself is increasingly difficult because of spiraling costs, people's unwillingness to move, and the short timeframe of many projects that does not justify the expense of moving people. Teams *must* spread out.

Distributed Teams

Distributed teams comprise people in the same organization who work in different places either interdependently (like a multisite product development group) or separately (like branches and local offices). Buckman Labs research and development operation is distributed across all the company's sites.

To carry out interdependent tasks, teams with members in different places clearly have a distance problem to solve. Perhaps just one person is situated remotely. Perhaps several are. Sometimes everyone is in a different place. More than a dozen people in Boise Cascade's Paper Division Printing and Converting Business group have moved out of the company's Portland, Oregon, offices—choosing instead to work from home.

The ability to work at a distance is reshaping the traditional headquarters-field relationship. Site managers belong to the same organization but rarely work as a team under the old model. Indeed, branch offices, one familiar example of de facto distributed teams, often are encouraged to compete in a system that pits one against the other in the effort to maximize output and beat quotas.

When branches work together, however, they form virtual teams of people in the same organization situated in different places. The 47 branch managers of BankBoston's First Community Bank formed a team among themselves to address common service issues and to develop cross-branch priorities based on their own needs and shared learning. Branch teams face the same problems of crossing space and time as more organizationally-diverse virtual teams. Local intergroup boundaries are sometimes more difficult to bridge than more distant affiliations because of competitive forces arrayed along contiguous organizational borders.

Even if not distributed in space, virtual teams can be spread out in time. Teams of shifts and groups of managers and professionals-on-the-move share facilities—people in the same organization who use the same place but at different times. Buckman's manufacturing plants run on three eight-hour shifts daily and, as Bob Buckman estimates, their management teams are in their offices just 14 percent of the time.

When Eastman Chemical Company began its long drive to develop teams in the late 1970s, they began with shift supervisors. As the interface between shifts became more complex, the information load increased, so Eastman formed teams within and across shifts that supported 24-hour quality improvements.

A new variation on the old time-shift theme is the galloping trend towards "officing." Flexible office arrangements include the "hoteling" concept, such as IBM's "sales warehouse offices." There people sign up for a workspace on arrival at the office. Other future-here-right-now work environments maximize collaborative workspace with mobile and home-based offices for independent work.

Distributed Cross-Organizational Teams

Distributed cross-organizational teams involve people from different organizations who work in different places. Buckman's Distance Learning Team of a dozen people from multiple organizations includes several who work from home, such as one member who lives in Boston and attends real-time Memphis meetings over the phone.

The classic virtual team combines people in different places and organizations with some need to function at the same time (synchronously)—not all of the time, of course. Most work combines a pattern of individual tasks and group tasks, time spent working alone and time spent working with others. For most virtual teams, synchronous interaction—shared time—is a scarce resource. Time together is planned, prepared for, and followed-up on. Hewlett-Packard's worldwide distributed product information management system (PIM System) team combines quarterly face-to-face meetings and extensive use of electronic media to function across global distances and 24-hour timeframes.

Time creates a complication that not even instantaneous communication can solve. As the distance increases and more time zones are crossed, the window of synchronicity in the work day narrows. New England is six hours behind Europe, and people in California leave work just as their counterparts start their next day in Japan. Even when real-time interaction is possible technically, it may not be practical.

The most extreme type of virtual team is one that is cross-organizational and that rarely and in some cases never meets in the course of doing its work. Without face-to-face time, this type of team tests the limits of dealing with contentious issues, but may shine for information-sharing and technical problem-solving tasks. Buckman's global conversation system provides conditions that allow worldwide cross-organizational teams to form within hours to work on a customer problem or opportunity that may last for days or weeks.

Variations on a Theme

Virtual teams also spring up in traditional settings. When a collocated team uses tools that a distributed team requires just to survive, they may achieve levels of performance far beyond the conventional norm. Buckman's many collocated teams have benefited enormously from their virtual team systems and the experiences of their colleagues around the world. Small collocated core teams can help support the needs of a larger distributed organization.

By implication, the traditional model of work meant that everyone in the group of course spoke the same language and took their nonverbal cues from the same broader culture. Today even when people are in the same location the chances of their speaking different languages are high.

Virtual teams break the traditional mold as people work across boundaries. When people occupy different places and come from different organizations, they can be certain that they will have to cope with difficulties in communicating across culture and custom with different languages. The language differences that virtual teams have to contend with are not all born of different country tongues. Two people from different professional upbringings can have almost as much problem communicating as two people who grew up speaking English and Japanese.

When teams go global, their language and culture issues clearly loom larger. However, all teams of the future will have to cope with the fact of increasing diversity in the workplace. Not only is the workforce becoming more diverse, but the task requirements of complex work demand that a more diverse group of people works together, whether in traditional settings or in virtual teams.

BankBoston's First Community Bank (FCB) tested the limits of the large-scale, cross-cultural, cross-organizational virtual team. Working in partnership with the Overseas Chinese Credit Guarantee Fund, a Taiwan-sponsored effort to make loans to Chinese business people around the world, Boston's FCB had to include players from many organizations. According to Steven Tromp, a director of commercial lending at FCB, the team included people from a dozen organizations within the bank, four people from Chinese community groups in Boston, and six representatives of the Taiwanese government. "And that was just the inner circle of the team," Tromp says. "Dozens more were touched by it as time went on including the business people who finally got the loans. It was all done through e-mail, phone calls, and 'who knows who.' It was a neat example of a virtual team."

Sometimes collocated teams have even greater difficulty than virtual teams dealing with variations of language and culture. Because they are less aware of their communication barriers, collocated teams do not necessarily create appropriate compensatory norms. There is an analogy here to the relationship between distance and collaboration. Data show that people are somewhat less likely to communicate with a colleague upstairs in the same building than with one in another building.[13] When people know they are at a distance—culturally and linguistically as well as spatially—they are more conscious of the need to be explicit and intentional about communication.

A System of Virtual Team Principles

The three-part model of the virtual team concept—people, purpose, and links—is a simple but powerful conceptual tool (Figure 2.4). With it, you can grasp something as slippery as a network and something as immediate as a small group. "People linking with purpose" is our

Figure 2.4   Three Foundation Concepts

"A small group is . . .

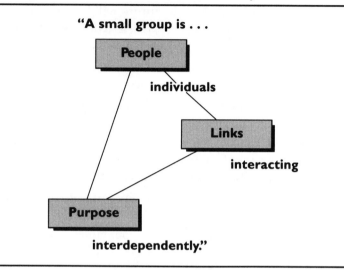

interdependently."

expression for the standard small group model "individuals interacting interdependently."

To account for the essential characteristics of virtual teams, we need to go down a level. In Chapter 1, we briefly outlined nine Virtual Team Principles that expand the three-part model, adding qualifiers to general team features that make them "virtual."

The nine principles together provide an integrated framework for understanding and working in virtual teams.

> *The principles of people, purpose, and links form a simple systems model of inputs, processes, and outputs.*

To start a virtual team, you need independent people, cooperative goals, and multiple media. As the team goes through its life cycle development process, people share leadership, undertake interdependent tasks, and engage in myriad boundary-crossing interactions. As the

Figure 2.5   Virtual Team System of Principles

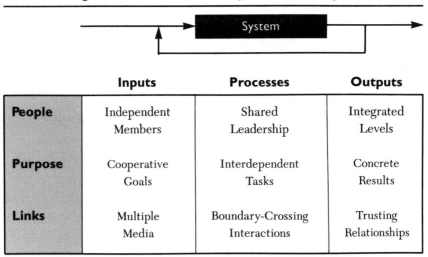

	Inputs	Processes	Outputs
People	Independent Members	Shared Leadership	Integrated Levels
Purpose	Cooperative Goals	Interdependent Tasks	Concrete Results
Links	Multiple Media	Boundary-Crossing Interactions	Trusting Relationships

team's life cycle unfolds, it produces concrete results, integrated levels of organization, and, if the teaming is done with integrity, trusting relationships (Figure 2.5). We describe these principles in greater depth in the following three chapters.

Virtual teams are teams and then some. Virtuality always adds a "spin"[14] to classic learnings about teams. The greatest difference between virtual and conventional teams is in their links. Within links, the explosion of electronic connections, particularly the new digital media, is the driving force of change (see Chapter 4). Virtual teams also include everything that is familiar about traditional teams, beginning with clear purpose.

CHAPTER 3

THE POWER OF PURPOSE
Do, Doing, Done

Purpose is the *sine qua non* of virtual teams. Nowhere is this clearer than at Eastman Chemical Company, which has taken a long but dramatic leap. An old-line industrial manufacturing company, Eastman is becoming a model 21st-century team-based network organization replete with virtual teams.

Eastman's Chemistry of Organization

As one of the key architects of its transformation, Dr. Robert Joines, Eastman's manufacturing vice president for Europe, the Middle East, and Africa, says:

> Eastman is a huge system of 800 to 900 interlocking teams that criss-crosses the company—and even includes our customers and suppliers. When we think of doing something, we think of teams. And, today, Eastman is almost a seamless organization.[1]

The time by car between any two of Tennessee's big cities—Memphis to Nashville, Nashville to Knoxville, Knoxville to Chattanooga—is under five hours. In our search for great virtual teams, we made the drive east from Nashville in the center of the state on I-40 on an early summer's night. The sun setting behind us colored the sky ahead, painting the peaks that become the Appalachians the shades of their names. Here, where Tennessee, Virginia, and North Carolina converge, the Blue Ridge meets the Smoky Mountains.

Smoky blue turned to black by the time we topped the last hills to our destination. Suddenly a surrealistic "whiteprint" of light flashed below us, shimmering strands of luminescence that sparkled for acres, pipes pushing up to 90-degree turns then angling down into elaborate twists of conduits. A spaceport? A city of the future? Tucked away in the small town of Kingsport is one of the world's largest chemical manufacturers—and an acknowledged world-class management system.

Spaceship Eastman

It was in this valley conveniently situated on both a navigable river and a railroad that, at the urging of a local entrepreneur, George Eastman decided in 1920 to invest in a defunct chemical plant. He needed a reliable chemical supplier for his developing photography business some 700 miles to the north, Eastman Kodak in Rochester, New York.

Three quarters of a century later, Eastman Chemical Company, which spun off from Eastman Kodak in 1994, is a $5 billion plus global operation with 17,500 employees manufacturing 400 products. Although you cannot go into the corner store and buy anything with the Eastman label, you would have a hard time not buying something with an Eastman product in it. Need a toothbrush? An Eastman chemical is an ingredient in the plastic handle. Soft drink and liquor bottles, pain killers, peanut butter, tires, carpet, mascara, stonewashed jeans, brake fluid, garden hoses, thermos bottles, latex paint, tennis balls, and, of course, Kodak film all contain Eastman chemicals.

Today Eastman Kodak is still Eastman Chemical Company's largest customer, but it is only one of 7,000 who buy its products around the world. Nestled in the northeast corner of Tennessee, Kingsport is the chemical company's world headquarters. Manufacturing operations stretch across six American states, Canada, Mexico, The Netherlands, Malaysia, Argentina, Wales, Hong Kong, and Spain. In 1993, Eastman became the first (and still only) chemical company to receive the U.S. government's highest kudo for quality, the Malcolm Baldrige National Quality Award.

We reported on Eastman in *The Age of the Network*.[2] There we told the story of a traditional firm that reinvented itself as a networked organization while still retaining important elements of hierarchy and bureaucracy. Since our first report, there have been some changes in Eastman's organization and a handful of "firsts" at the company. For example, as a result of the spin-off from Kodak in January 1994, Eastman must now deal with investors for the first time. With its global presence growing rapidly (the goal is 50 percent of its sales from markets outside the United States by the year 2000), the company has had to strengthen its geographic links. For the first time, it named regional presidents in Europe, Asia Pacific, and the Americas.

In this book, we explore the virtual teams fueling Eastman's transformation.

Teams Every Which Way

It took a quality crisis of business threatening proportions—and nearly two decades of work—to move the company to the structure it now has.[3] The company began its renaissance in the late 1970s when it lost market share of a major product due to poor quality. With a focus on traditional quality approaches and a lot of common sense and creativity, Eastman went about redesigning how it did its work from the shop floor to the very top of the company. Today, *everyone* works in teams and usually in multiple teams.

Ask people at Eastman why they won the Baldrige Award and they will point to their quality philosophy that rests on team alignment. "It's a consensus style of management," says Earnest Deavenport, Eastman's CEO since 1989, "that is much more based on team than individual decisions. There is more empowerment of teams than you find in a conventional organization. Many fewer decisions get bumped up to me to make individually."

Eastman is a complex mix of permanent, temporary, face-to-face, ad hoc, geographically distributed, culturally diverse, vertical, and horizontal teams. Some have traditional team leaders. Some rotate leadership. Some are quite formally chartered. Others less so. There are multiple executive teams and hundreds of shop floor teams.

Almost all Eastman teams cut across space, time, and/or organization boundaries.

Eastman has all types of virtual teams. It has shift teams that are responsible for keeping operations going around the clock. Short-term project teams are invariably cross-functional. While sometimes collocated, more often these teams follow the typical pattern of coming together and then going apart (see Chapter 6). They meet as necessary to plan and align their work then carry it out individually or in smaller groups.

The presidents of all the manufacturing facilities comprise a self-directing executive-level virtual team. They are members of the same organization with common problems and responsibilities situated in different locations. They rotate the chair every quarter. "We had teams working in the operational level, but somehow it seemed we didn't trust plant managers. We did away with that and said, 'Why don't you guys work together to manage manufacturing for the whole world?'" comments Deavenport.

Longer lasting process teams as well as customer and supplier teams are distributed and cross-organizational. Although most virtual teams need some face-to-face time together to function effectively, especially at the beginning, they can become "more virtual" over time. Eastman has a supplier team, for example, that had many face-to-face encounters when it began but increasingly fewer as time passed. Once the group established trust and set up its processes of interaction, it continued to make quality improvements without meeting. The virtual team functioned asynchronously in different places and organizations.

As it opens new chapters in its teaming journey, the company has begun to look at *time* as a major resource. "Everyone has the same amount of time available—24 hours a day," observes Lynda Popwell, Eastman's vice president of Quality and Health, Safety and Environment (QHSE). "We want to make sure that everyone in the company is focused on creating value for the company in everything that we do. Period. We want to reduce the complexity and to work on only those things that really add value to our customers."

At one point, Popwell stopped to count how many teams she was on and found it was too many. "I had been invited to be on a number of teams and I had accepted these offers until all of a sudden I realized it was too many. Then you have to assess and use good quality management principles to figure out what your priorities are."

Getting Quality Together

Because Eastman is a manufacturing operation, good teamwork on the shop floor is synonymous with good quality in its products. One of its earliest "improvements" came as a simple recognition in the late 1970s: the four shifts per day were simply one ongoing team spread out over 24 hours.

"At that time, we had a 'tag and you're it' shift change mentality, four different people around the clock running four different shift teams," says Will Hutsell, Eastman's senior associate in Corporate Quality. Very little information passed between shifts. When one shift left, the next would come in and readjust all the control room dials, as if the people before them had no idea what they were doing. Operators had to ask permission to make any changes and, of course, they punched time clocks.

During the initial implementation of continuous improvement efforts, Eastman took groups of the shift foremen off the job for quality management training. Instead of working their normal shifts, they came together as a group for training and planning improvement projects. These foremen (who are now called team managers) would then go back to their work areas and hold team meetings with their operators to develop plans for implementing the projects.

"Before long, the control rooms were transformed," Popwell explains. "The operators had the skills and information they needed to do their work without asking anyone's permission. In time, they stopped punching time clocks. Operations improved enormously and control rooms were clean." Eastman was on its way to a Baldrige.

Today, Eastman's training process is a sophisticated operation that reaches across the company. For example, to be trained as coaches, people leave their jobs for 14 weeks of intensive education in modern management thinking, skill building, and practice. Initially, they set up training programs and had people attend individually, but they found that approach not to be very effective. So groups of managers started attending together, creating sufficient critical mass of shared experience in the organization to sustain the learning when people returned to their jobs.

At the same time that Eastman reinvented work on the shop floor, it also made many other significant changes that raised the trust level in the company:

- It equalized benefits across the organization. Everyone has the same vacations for years served and everyone has access to the same healthcare plan.
- The executive lunch club became the business dining room, open to anyone with a business need.
- It eliminated the traditional performance appraisal system that distributed people's performance across a bell curve and replaced it with an employee development system.
- After experimenting with team rewards, it stopped them because it proved impossible to draw indisputable lines around "the team." Individual teams depend on the overall interteam environment; they cannot succeed without it. Eastman now has a companywide bonus program.

Beginning with Purpose

All Eastman teams have a vision and a mission and most have charters and sponsors. In many cases, teams accept a written charter with a signing ceremony that commissions the teams.

"Because there are many teams at Eastman, it is essential that we always define the purpose of each team," Hutsell says. Without clear purpose and an established process for defining it, confusion not quality would rule at Eastman.

"Typically a broad charter is put in place and the team is empowered to see if it makes sense," he says. "The team can modify its own makeup." Using Eastman's Quality Management Process, teams pay attention to questions such as Do we have the right purpose? and Do we have the right membership? This helps keep teams on track.

The newly formed Latin America Team, for example, came together for several days to define its vision, goals, key results, roles, and responsibilities. With the purpose set, the team continues to meet but not everyone needs to attend every meeting.

"You must look at the purpose," Hutsell cautions. "Only when you have that right can you get from here to there." Eastman's early attempt to use one of the most fundamental quality tools, Statistical Process Control, a quantitative method for improving quality, failed when the company tried to implement it without fully making clear its connection to purpose.

From Intent to Results

The best predictor of virtual team success is the clarity of its purpose and the participatory process by which the group achieves it. Eastman teams thrive in a culture infused with purpose.

Purpose starts at the top, but is not dictated by the top. The Senior Management Team is the keeper of the Strategic Intent. This document contains both the company's vision, "To be the world's preferred chemical company," and its mission, "To create superior value for customers, employees, investors, suppliers, and publics." The Senior Management

Team is also custodian of the document known as "The Eastman Way" that sets out its values and principles. It emerged from an intense mid-1980s internal review of trust and barriers to team success. The document makes explicit the culture required to support teams working across organizational boundaries. It also serves as the touchstone for innumerable changes that make a difference in the day-to-day working life of Eastman employee-owners.

While written down and published, the Strategic Intent is also a dynamic document. It serves as the framework for overall long-term company strategy and the shorter term initiatives known as MIOs, Major Improvement Opportunities. Everyone in the company is involved in the planning process. Organizations bring the knowledge of their part of the business—specifically including customer needs, competitive comparisons, company and supplier capabilities, and risk assessments—together with the Strategic Intent to develop Strategic Alternatives. These are then formally submitted to the Eastman Executive Team that forms the overall strategy and makes selections among alternatives.

With the strategies set, the organizations develop goals, their own MIOs, and their criteria for success, known as the key results areas. With concrete measures in hand, the virtual teams develop plans using Eastman's Quality Management Process in which everyone in the company is trained. Eastman's version of the familiar "plan-do-check-act" continuous improvement cycle adds a step at the beginning. Before "plan," there is an "assess organization" step where the team focuses on clarifying its purpose in interaction with customers and in alignment with the company's strategy.

Early in its team journey, Eastman made a mistake that many companies do: They convened teams and focused on the mechanics of the meetings such as following agendas and drafting plans without giving the teams a clear purpose. "We learned how important it was for each team to understand its relationship to improvement projects and improved business results. Without this understanding and sense of purpose, teams did not accomplish what they were capable of accomplishing," Popwell says. That turned around when the teams took on specific projects with clearly defined expectations. Eastman values time spent gathering information, developing and choosing among alternatives, creating implementation plans, monitoring and documenting progress, celebrating results, and even formally disbanding. The process itself generates continuously increasing value.

Pass the Pizza

For people interested in organizations, Eastman is famous for its unique organization chart, its "pizza with pepperoni" depiction (see *The Age of the Network* for the detailed chart[4]). Hierarchy, bureaucracy, networks, and teams all have their place in this organization-in-the-round (Figure 3.1). The company expects people to communicate horizontally not vertically. There is no going up a chain of command, across to another

Figure 3.1   Eastman Organizational Design

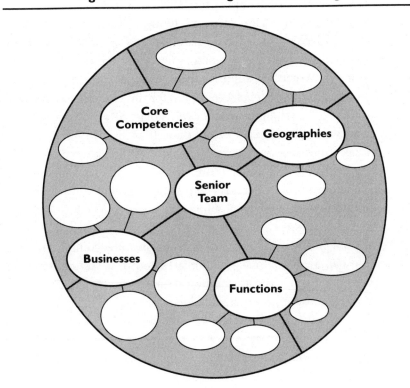

function, and then coming back down another chain of command to get something done.

Eastman is organized in four vectors: Businesses, such as Coatings, Inks, and Resins; geographies (Worldwide Business Support); competencies, like Polymer Technology; and functions, such as manufacturing.

Deavenport created the circular chart in 1991, he says, because he wanted "to signal the organization that this was a different structure, a networked structure, a team effort, not business as usual. We're all in this together." While the chart preserves the logic of hierarchical levels, "the artist in you has to come out to see the pizza chart as different from a hierarchical one," Deavenport comments.

Playing Out Purpose

"We develop a yearly business plan for HSE (Health, Safety, and Environment) that links my organization to the overall company strategy," explains Popwell. "We gather inputs from business organizations, manufacturing sites, communities, regulatory agencies, audits, assessments, and benchmarks. We conduct gap analyses using our key results areas. Using all these inputs, we develop our plan for the year, establishing the initiatives for HSE at the corporate level, along with the specific supporting plans for achieving the overall plan. It's a living document that we modify as circumstances change. This process is repeated by my direct reports; it establishes linkage throughout the company and ensures that everyone clearly understands the purpose of what they are doing. This is *my* pizza chart," she says, pointing to a page in the business plan with a swirl of circles that describes her organization.

"We use our business plan to know where we are going," she says. "We can do gap analyses to see where we need to make adjustments. All Eastman organizations use a similar planning process to align with corporate goals and with their stakeholders."

Purpose is the metaphorical campfire around which members of the virtual team gather.

Purpose generates the internal spark of life for task-oriented, boundary-crossing virtual teams. To survive, they must turn their purpose into action, using it to design their work and organization. Some teams receive a fully formed charter; others go off with a vague sense of desired direction. Some team members think that setting purpose is important while others do not. Most team experience lies between these extremes as people struggle to understand and express their purpose.

Abstract to Concrete

We use the word "purpose" to stand for a range of terms from the abstract to the concrete—from *vision, mission,* and *goals* to *tasks* and *results.* The essential first step toward making purpose useful is to untangle these concepts. By ordering these terms and relating them to one another, it is easy to adapt this framework to the language of any organization. For example, some call visions "guiding philosophies," missions "charters," goals "objectives," tasks "activities," and results "outcomes." Regardless of which words you use, the important point is that they all interrelate. Purpose is a system of ideas that expresses itself differently across the organization yet carries common threads.

Improving Your Vision

In the organization driven by purpose, vision is the font of inspiration, the source that generates the flow of work. When articulated best, visions include a compelling picture of an achievable, highly desirable future. Vision is also the realm of values and philosophy, the intangible but crucial culture of ethics, norms, and the intrinsic value of people that bring life to virtual teams. It does not have to be long. Eastman's vision is short and simple: "To Be the World's Preferred Chemical Company." The word "preferred" carries both the customer focus and the implied strategy of superior quality to achieve the global goal.

Other organizations create lengthy vision statements. The Massachusetts Teachers Association, with 75,000 members—one of the largest unions in the Commonwealth—has a vision statement that covers a page.

Mission (Im)Possible

Mission is the simple statement of what the group does, its reason for being, expressing its identity. Though usually more specific than the vision, it is still quite abstract in its attempt to capture the essence of the organization in a handful of words. One kind of mission statement answers marketing expert Ted Levitt's famous recurring question, "What business are you in?" In a world of fast-changing markets, this question usually goes to the heart of organizational transformation and the teams it spawns.

Virtual teams need particular clarity around vision and mission, setting a broader context for details of work because routine procedures and policies are not available. This is true even when the members of a virtual team all come from rigid bureaucracies. The Office of Law Enforcement Technology Commercialization project, a virtual team funded by the National Institute of Justice, expresses its mission concretely, "To assist private sector entities to bring law enforcement technologies developed in federal laboratories to market."[5]

"Your Goal—The Sky"[6]

Goals stand at the midpoint between intangible lofty visions and tangible, practical results. They reformulate the mission—the singular overall goal—into a handful of doable subgoals. Like the population of the groups that think them up, goals are most effective when they are few in number. Goals provide motivation. They are the starting point for work processes, the original way to divide the work into its components. Goals allow you to gaze into the future at the desired outcomes. Embedded in the selected set of goals is a strategy for how vision and mission are going to turn into positive results. Smart goals derive from clearly considered strategies, an effort that may comprise the greater part of the planning process.

When a new international software organization held its first global planning meeting, the group identified 10 cross-cutting goals, then selected three as strategic priorities. Within a year's time, they had substantially accomplished each of the priority goals and established themselves as a business unit.

Task Masters

Tasks are "how" the work is done, the actions that arise as pieces of work that put goals into action. Invariably expressed as verbs, tasks are the specific actions that team members take. Tasks are the signature of teamwork, the very seed of the definition of teams. Purpose becomes quite practical when expressed as tasks. Because they are constantly in motion, tasks can be somewhat slippery to the materialist grasp.

Minnesota's Department of Natural Resources, which uses virtual teams extensively, employs a 15-member team to integrate its planning and budgeting processes. "We get together about once a month as an entire group and subdivide the tasks so that people who are more geographically contiguous can work together," says Terri Yearwood who is responsible for the department's overall strategic planning process. "We also make sure we have the right mix of skills for the task." By subdividing the work, more people can contribute to the results.

Nothing Succeeds Like Results

Results thud into place as the concrete outcomes of purpose. Reports, presentations, events, products, decisions—outcomes that everyone understands—clearly express purpose. The team creates *something* through its work. For a task action to be complete, there is always a result—however grand or poor—within a given period of time. It is in the nature of the task-oriented team to produce and judge itself by its end products.

For *Men's Health, Rolling Stone,* and *Esquire* magazines (see Chapter 1), the concrete result of their virtual team's work was a proposal to their client that won them all lucrative contracts.

Purpose Seeks Results

There is a natural flow to the levels of purpose. Like mountain streams seeking springs, purpose runs from the heights of abstract vision to concrete valleys of results. Putting purpose into action over time produces a dynamic picture of work (Figure 3.2).

Vision flows into mission that becomes the highest level goal. To accomplish the overall purpose, mission segments into subgoals as the first

Figure 3.2   The Flow of Purpose

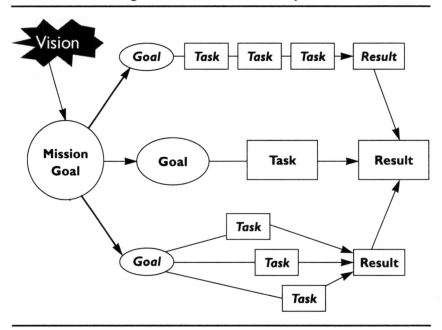

step toward the division of labor. Tasks are the specific steps taken to achieve results. Some steps are serial, each dependent on the one before. Other task sequences are independent and parallel, coming together at the end. Some tasks are simple and lead directly to results; others are more complex. Most work undertaken by a virtual team will mix all these forms.

Purpose as a Path

To put purpose into motion, plan. The better the planning, the more effective the process. Better processes have a satisfying pay off: They require less time to fulfill the purpose. Many groan at the thought of planning, harking back to innumerable experiences of plans that were shelved, contradicted by higher authorities, perhaps detailed to a level of mind-numbing minutiae. As bad as planning can be, for virtual teams there is no other way. They must clarify and refine purpose into the process of work to accomplish results. This means planning.

We naturally know how to plan in a general sort of way—and do so all the time. The design of work for virtual teams follows a universal pattern of behavior that we use repeatedly every day—the path. Each time we envision a result that we act upon, we conjure up a little bit of planning, whether to go to the kitchen to get a cup of coffee, to the store to get a magazine, or to the garage to get the car fixed. We have an image of what we want in our minds (a goal) and we rehearse the steps (planning the tasks) that will take us to get what we want (the result).

Goal → Tasks → Result

These interrelated terms reflect the universal pattern of work: a motivating source, a target at the end, and a sequence of steps that connects the source and the target over time.

Short paths or long journeys, the logic is the same. Most of the journey lies in the middle, between goals and results, in the province of tasks, the doing. While tasks are the path itself, the pieces of work that fit together over time, process is the sequence of actions. Tasks are verbs—for example, plan, develop, decide, negotiate, do, write, deliver. Each task is itself a little bundle of *cause-and-effect*. Results do not just appear by magic at the end; they grow over time in the course of doing work, performing tasks. Tasks produce interim results, parts of the final result.

Corporate Breakdown

Virtual teams truly are microcosms of the organizations that spawn them. This becomes increasingly evident the higher the group is in the organization. At the top, the senior team most literally and directly expresses that truth. There all the work components come together, and the organizations of all the major players cross boundaries.

Put into the language of purpose, the structure of hierarchy and bureaucracy can be seen as a way to break down the work of the company as a whole into interrelated chunks.

Look to the interplay between goals and tasks to see this little recognized but profound fact of business life: Goals at one level generate tasks at the next level, where the assignment is in turn taken on as a goal. Vertical and horizontal work process design is a very natural and practical way to organize teamwork at every level of an organization. Although it is not a conventional way to view corporate structure, the goal-driven work process offers teams a way to look at how their work interrelates across the company (Figure 3.3). The purpose-oriented approach to organizational design is essential for extensive virtual teamwork.

The board of directors, representing the owners, sits at the top of the organization together with the highest ranks of senior management. In most business organizations, senior management generates the vision and provides direction, whether bold or timid, vivid or obscure, conscious or unconscious. The hierarchy of goals is established based on what the owners want: profit, growth, and other returns important to them. The corporate goal is to get these bottom-line results.

To reach positive corporate results, the company creates value through its work. Most organizations (save those which operate on a completely ad hoc basis) divide the work as a whole—literally—into divisions and departments (both derive from the same root word meaning *divide*). Each has its own mission—for example, marketing, research and development, engineering, production, sales, and distribution departments (functions) or divisions based on factors such as products, geographies, or industries. To make its contribution to the joint work effort, a large organizational component like marketing may further subdivide its piece. Internally, it creates its own departments and groups, each of which has separate charters (written or "just understood") based on the work assigned from above. A sequence of interrelated positional teams is created at every level with their own work to do that carries its expected outcomes.

The conscious formal design of an organization in this way is just what Eastman means by its "vertically interlocking teams." Eastman's

Figure 3.3 Corporate Purpose Breakdown

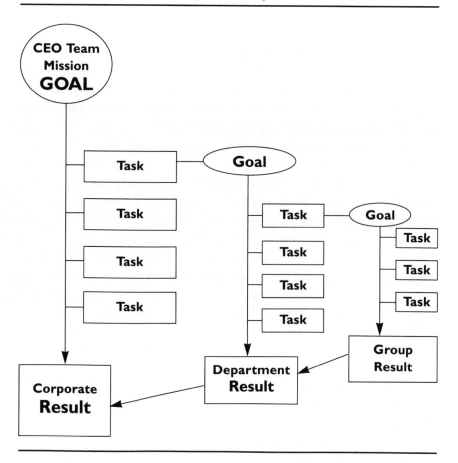

permanent teams start at the top with the executive team. "This is where the interlocking functional teams begin, the vertical ones. You become a member of the team because of your job function," explains Popwell. "I'm on the executive team because of my function. The executive team is Earnie's [CEO Earnest Deavenport] team and it sets company strategy. Everyone on Earnie's team has their own team. I have the directors of health, safety, product safety and regulatory compliance, environment, and legal representation on my team."

So it goes down the line, with a person from this level serving on a cross-functional team while also serving as head of the team at the next level down.

The vertical team structure that Eastman makes explicit exists in most organizations whether or not they recognize it. Many executive and management groups function as staff rather than behaving as teams with consciously interdependent tasks and an explicit responsibility to optimize across the larger organization. Behaving as staff has its role but executives can be even more effective when they function as *teams*. Ideally, the formal organization design would perfectly reflect the intended markets and the products and services that the company produces. Today's perfect form is tomorrow's mold to break as markets shift, technologies change, and knowledge advances.

While vertical teams are the functional, geographic, and product components, the horizontal teams are where they work on *cross*-organizational processes. "Processes are how the work gets done. These teams deal with key issues," Popwell says. For example, Eastman's vice chairman and executive vice president Wiley Bourne's team consists of the new regional presidents in the geographies, the hierarchy of the team. At right angles to this team are teams that deal with specific issues such as the Latin American Growth Team with members in Mexico City and functional people in Kingsport and in Argentina.

Eastman's vertical interlocking teams are the most permanent, but some cross-functional process teams are also quite permanent, such as the Order Fulfillment Team, the Eastman Training Process Team, the New Business Generation Team, and the Innovation Team, a multilevel team that involves people from every major area in the company including technical service, the business organizations, sales, and research and development.

Innovation is part of a key group of horizontal teams centered on the company's seven core competencies, which it regards as the foundation for the company's Strategic Intent. "It is the responsibility of core competency teams to make sure you have the skills and hardware to do the job," says John Steele, Eastman's director of Technology Core Competency. Within three scientific competency areas, 26 technical "engines" have been identified, each with a technology leader and team of experts.

"[In polymers] we have a stake in polyesters so that engine is polycondensation where we are trying to judge whether we have the pilot plants, labs, skills, and knowledge to meet business needs."

Detailed Digital Vision

While the importance of clarifying purpose does not by itself distinguish virtual from conventional teams, the depth and clarity of its expression does. The purpose problem is two-fold for virtual teams:

- Crossing boundaries of space, time, and organization only further complicates already complex communication. An inherently messy process of creating a coherent, productive, and lasting purpose in the early stages of a team's life is even less tidy for a virtual team. It needs dense and frequent communication.
- Once developed, the team must make the purpose and plan explicit in symbols, words, diagrams, tools, and handbooks. The plan also must stay updated, flexible, and adaptable in order to serve as a coordination hub for distributed work.

The problem with a complex purpose is people's need to grasp it in simple ways. Establishing purpose is first a conceptual problem, then a display problem, and finally a navigation problem.

"Periodically, we go back to our original purpose, our shared mental model. Building such a model has been extremely helpful in communicating across geographies and cultures. Some teams produce explicit pictures of what their mental models have become in terms of numbers and graphics. It gives us a vision of where we are headed that allows us to plan for what we need in terms of specifics such as logistics, sales, and technical service," Eastman's Hutsell says.

Interactive digital media offer a wealth of untapped potential for virtual teams to expand their communication capacity. At the same time, the expression and direct use of the power of purpose really comes into its own with the World Wide Web and intranets.

With the Web, mental models, whether expressed as outlines, lists, diagrams, or art, can easily be displayed for and used by the team and its partners. These mental models can then become portals—quite literally as "clickable" links and maps—to layers of more detail about goals, tasks, results, people, resources, organization, and every other kind of information that may be important to a team's work (see Chapters 7 and 8).

Mental models—ranging from broad perspectives on the market, estimates of risk, and organizational strategies to budgets, product designs, work processes, and agendas—have been the province of hierarchy. Typically, they are locked in the boss's head or file drawer, developed through experience, and communicated to others as needed. This works in the world of slow pace and simple purpose where people are not expected to think for themselves. But fast-paced virtual teams facing complex problems need to share a conceptual framework of their work. With the advent of digital electronic media, they have a powerful new communications tool that brings purpose alive.

In the networked world of the Information Age, purpose is the new source of legitimacy and power.

The Authority of Purpose

Nothing is more important to the virtual team than a clear sense of purpose. Hierarchical groups can fall back on force as their source of authority. Bureaucracy can turn to rules and regulations. Virtual teams require something more to mold them and hold them together. When new forms of organization emerge, so do new sources for developing cooperative social structures, new premises of authority and power. To see how authority works in virtual teams in the age of the network, we need to start at the beginning.

Four Ages of Authority

The ability to demonstrate competency, inspire passion, and recall the voices of the ancestors were the foundations for authority and power in the Nomadic Era. Leaders in traditional nomadic cultures (including those surviving today) had the ability to influence people but generally not to force them to do anything.

Charisma and tradition ruled nomadic cultures.

Eventually nomads put down permanent roots to tend farms and build cities. Hierarchy arose.

Hierarchies use force to defend resources, maintain social stability, and control technology.

In business, the people who sit at the top of the hierarchy—the owners—have the power to hire and fire, to give out rewards, and to inflict punishment. They may promote and demote employees, grant and refuse raises, acknowledge people or place them on probation. The extraordinary power of owners is nowhere more evident than in their exclusive right to buy and sell a business as a whole or in parts.

When agriculture gave way to industry, brute force yielded to the rule of law.

Bureaucracies use law and its derivatives—rules, regulations, policies, and procedures—as the fundamental basis for authority and legitimacy. Bureaucrats take their authority from their place in the administrative structure, drawing on the legal perks of their positions.

In most organizations, people's positions reflect both the reward and punishment power of hierarchical rank with the resource control power of bureaucratic specialty. As Eastman's vice president (hierarchy) of Quality and Health, Safety and Environment (bureaucracy), Popwell has the formal responsibilities of both her rank and her function.

Purpose has always been important if not central to small groups and teams. In the information era, it takes on a new aura as the source of legitimacy itself. The legitimacy conferred by jointly held purpose is uniquely vital to virtual teams. Because of the diminished role of traditional authority, they need some other guiding force.

Virtual teams develop an inner authority based on their members' commitment to shared purpose.

Strategic alliance teams, for example, consist of people from different companies. They lack a common hierarchy or set of administrative policies since the employees who populate them have totally different reporting structures. In cross-functional teams, perhaps the best-known type of virtual team, no common authority figure may tie them together until they reach the CEO. Such is the case with Intel's Native Signal Processing project, a technology breakthrough requiring the efforts of dozens of groups and literally thousands of people throughout the company whose only common reporting structure comes together at CEO Andrew Grove.

Knowledge Power

Inside or outside formal channels, within a company or between firms, people with an idea inspire or recruit others to join them and a virtual team forms with purpose as the essential glue. Ad hoc teams self-legitimatize through common purpose.

New forms of authority exist long before they become dominant. Law emerged in the hands of Hammurabi and Solon thousands of years before the rise of constitutional governments like the United States that submitted military force to democratic rule. From laws, rules, and regulations, industrial bureaucracies constructed newly dominant modes of authority.

Just as each age strengthens a new source of authority, so does it bring a new basis for the other half of the governance equation—power.

> *In virtual teams, power comes from information, expertise, and knowledge, the new foundations of wealth.*

Purpose defines the work that, when reduced to its parts, becomes tasks. Specified tasks require specific expertise. Experts have always had power but not the kind that they do today. Peter Drucker regards knowledge as so critical to people in the emerging information economy that he uses the word in the plural—we need multiple "knowledges" to survive.[7]

People with common purpose work out a cooperative set of goals. They come together with the hope that by combining efforts they can achieve great results. To get to results, they naturally divide the work into tasks that require people with specific skills, capabilities, and experiences. People with needed expertise and knowledge have power to the degree to which the work requires them (Figure 3.4).

Virtual teams have an especially tough job. They need to cope with all the traditional sources of power and they must harness the new forces of information and knowledge power.

- Though it remains unclear how organizations will adapt the reward and punishment systems of hierarchy to virtual teams, hierarchy clearly has far less influence in the new world of virtual work.
- The importance of position in virtual teams varies enormously. Some individuals come to the virtual team as anointed leaders. In other situations, everyone temporarily (and sometimes uneasily) sheds their rank and takes on multiple new identities. Formal positions still exist, but they often are not determining factors in virtual teams where the structure of the work takes precedence.
- Because of their ability to reach anywhere for members, virtual teams can easily include people with a wide diversity of knowledge and skills. By deliberately seeking difference, the team

Figure 3.4   Sources of Authority and Power

	Source of Authority	Source of Power
Individual	Charisma	Personal
Small Group	Tradition	Affiliation
Hierarchy	Ownership	Reward/Punish
Bureaucracy	Law	Position
Network	Purpose	Knowledge
	↓	↓
	Goal-Based Authority	**Task-Based Power**

reaps the creative benefits of a broader range of viewpoints and expertises.

Everyone on a virtual team is or should be expert in something needed for the group to accomplish its work. The more important the work, the more highly valued are the required skills. Like architects, consultants (who represent relatively pure instances of people with expert-based power) build teams to work on a specific project then build another team to satisfy the needs of the next project. Each team composition differs depending on its requirements.

Rewarding People for Success Together

"Just look across the street," Eastman's Popwell instructs, pointing toward the company's huge coal gasification project. "This state-of-the-art facility produces chemicals from coal, using advanced environmental controls. At one time, there was a lack of communication between those two departments—Gasification and Acetic Anhydride. They worked as a team across department lines to solve the problem, resulting in a savings of $1.5 million. This is a good example of synergy that results in a better solution."

With a common purpose, the two teams established strong links. There is a very strong incentive for Eastman teams to come up with ideas that save the company money. The more profitable the company is, the higher everyone's bonus.

In the first year after they went public—what they call "the spin"— Eastman instituted one more new thing that they jokingly say is "much better than the ham biscuits," a traditional part of reward events: the Eastman Performance Plan, a *companywide reward*.

The payout in 1995 was significant: Everyone across the company got a 30 percent bonus.

"One of our goals is for everyone to be owners in the company," Popwell says. Through a 5 percent salary give-back, employees created a base pool. The first 5 percent of the annual payout then comes back to them in an employee stock ownership program. "This makes us all owners and managers and that encourages Eastman people to make good business decisions at every level," she says.

The payout is based on return on capital minus the cost of capital. In the first year after the spin, the payout was 17 percent. "The Performance Plan rewards everyone at Eastman—the big team—based on company results. We have teams that link across the company, vertically building on our interlocking team structure. We have horizontal teams that manage and improve processes that cut across organizations. And we have teams that span the global regions around the world," Popwell says. "All these teams are focused on the corporate strategy and goals, because we don't want to optimize one area and jeopardize another area. Our employees are very interested in the well-being of our company. You often hear people say, 'This will affect our payout.'"

"Why do we have teams?" Popwell asks. "It's not because they are stylish. If you teach people to reach a level of understanding and goal sharing, you can get beyond consensus. You come out with a solution that is much better than any one individual could have come up with. That takes you to a higher growth level which is what you need for major improvements. The why of having teams is fantastic!

"Don't let us con you into thinking this has been easy," she warns. "It's been a difficult journey, and everything is not a success story. We have had to learn quite a bit as we've gone along. The reason we've been more successful as we've become larger and more global is because we have had the cross-functional team ethic. We're used to working in teams."

CHAPTER 4

THROUGH THE WORM HOLE

Links for Virtual Teams

Telecommunications and the global economy have arrived for millions of people, bringing with them new partners in daily work. Colleagues can sleep in opposing time zones and still be members of the same team. This is true, but it is not easy. The explosion of links across every conceivable boundary is staggering in its complexity as languages, cultures, governments, distance, and the mysterious nuances of human behavior all play their part.

The Once and Future NCR

"Hey, Gene. I've been meaning to talk to you," Kathy Black, a computer engineer for NCR Corporation, called out, walking into the conference room. "Is this a good time?"

"Just great," her colleague, Gene Young, a computer architect, replied. "I need to talk to you, too. Have a seat."

Kathy, the manager of Scaleable Systems for the project, spotted Gene, an NCR Fellow, "sitting there" as she was going down the hall to get a cup of coffee. Nothing unusual about this scene—one person needs to talk to another, asks if this is a convenient time, and sits down. Then why are the words "sitting there" in quotation marks? Because Kathy was in San Diego, California, and Gene was in Columbia, South Carolina, and suddenly they were both sitting down at the same table. Is this possible?

Not literally, but almost. Kathy and Gene, both members of the same team, worked together on a daily basis even though a continent separated them. They discussed strategy, argued points, solved problems, made presentations, exchanged documents, used flip charts, and shared files. Nor did they work only with each other. Their project colleagues numbered more than 1000 who worked for more than 11 months in three locations (including Naperville, Illinois) to develop a next-generation computer system.

The three-site virtual team was connected by a high-speed, full-bandwidth continuously available audio/video/data link[1] that they affectionately nicknamed "the Worm Hole." The Worm Hole—think of it as "a portal of instant transport from one place in the universe to another"—comes from the *Star Trek* TV show *Deep Space Nine*, which suggests such an intergalactic phenomenon in its opening credits sequence.

Making Its Mark with WorldMark

Kathy and Gene and their colleagues were all members of Dayton, Ohio, based NCR's new product development team for its WorldMark™ line of enterprise computer servers. These servers are today what mainframe computers used to be—places that house massive amounts of data and shared software made available to myriad individual users (known in the computer trade as clients).

The WorldMark development process accomplished something that few technology projects do: It met the market four months ahead of schedule. (So frequent are delays in technology development that there is no commonly used word to describe a project which does the opposite of "slip.") WorldMark is a great example of a globally distributed, cross-organizational virtual team. Guided by a clear purpose, the team used the most advanced communication links that we have encountered in the course of our research.

WorldMark's product family spanned, rather than simply filled, major market segments. The product can scale from relatively small computer configurations that link a few processors (the chips that are the computer's brains) to huge ones—very large scale massively parallel processors. Such behemoths can only be meaningfully described to lay people by the weight of the disks they use: In 1996, its 11 terabyte (a million megabytes) version weighed 20 tons. Such "terabrutes," as the NCR people jokingly call them, are used in organizations that manage massive amounts of data, such as banks, large retailers, telecommunications companies, and other organizations with global data infrastructures.

For NCR, the development of the WorldMark line turned into an epic project for the rebirth of the company. Founded in 1884 as the National Cash Register Company, the maker of the first mechanical point-of-sale devices initially got into the computer business in 1952. AT&T acquired the company in 1991, renaming it AT&T Global Information Solutions (GIS) in 1994. When AT&T announced its decision to break up into three separate companies in 1995, it named GIS as one of the units to be spun off. It would become an independent publicly traded unit—as the "once and future" NCR. While renewed independence was appealing to NCR and its newly named chairman and CEO Lars Nyberg, its corporate challenge was considerable. With 1995 losses of $722 million, NCR was looking to WorldMark along with several other initiatives to help return the company to growth and profitability.

It did. By the second quarter of 1996, NCR returned to profitability, reporting operating income of $29 million by the third quarter. The company appeared to be on the upswing. WorldMark's expedited entry into the highly competitive computer server market was a significant contributor to increased revenues. It represented both a process and a product success.

Instant Communication Through the Worm Hole

"We used various communications mechanisms to keep this very, very far flung team together," says Dennis Roberson, chief technology officer and NCR senior vice president. "The activities between San Diego,

Columbia, and Naperville in particular were kept together through the Worm Hole."

Roberson is a veteran of 25 years of doing projects involving people situated in geographically separate locations. "Misunderstandings are always a part of it," he says. "They are something you have to work very hard to fix from a management standpoint. In projects like these, there is always a characteristic lack of trust, particularly when you have groups in different time zones. The boundary lines between groups are not always clear. You've got existing groups and existing emotions and the challenge is to make everyone feel like part of the same organization."

One solution was the Worm Hole, which is, in Roberson's words, "video conferencing taken to its logical next step. It is a continuously open lease line so that you can have a meeting any time you want to. With the high performance of the link, it's just like being there." Because of the quality of the telecommunications connection, there was "no strobe light effect and the sometimes not so humorous delays," Roberson says. He refers to the annoying aspects that people associate with what he dubs "traditional" video conferencing. "All of that was overcome by the technology and great bandwidth," he explains.

"We did some nice things with the link between Columbia and San Diego. The grain of the wood on the table is the same in both sites so it looked the same," he recalls. By angling the cameras properly—there were two or three in each location along with 32-inch television screens—the desk in one location blends right into the desk in the other location. "It was just a bright engineer who thought of that level of detail," Roberson remembers. "That sort of thing really helps create the feeling of 'being there.' It's the only room of its type that I've experienced where you really do forget that you're not in the same place."

This proved so true to the group that it created its own standing joke. With its three-hour coast-to-coast time difference, one group always seemed to be having lunch while the other was not. "Someone was always saying, 'Can I pass you a sandwich?' People thought they should because they felt like they were in the same room. People couldn't help but make the offer and then laugh about it."

Each of the screens in the Worm Hole served a different purpose. One showed the people at other end. A second was the equivalent of an overhead projector that electronically projects foils onto the screen. The third was a standard PC monitor that facilitated information sharing and distribution.

As important as the Worm Hole was for organized meetings—the system could accommodate up to three sites simultaneously with as many people at each site as could comfortably fit into the 18-by-24-foot conference rooms—"the next step is even more amazing and more desirable," Roberson says. "When the rooms are not being used for meetings, the doors are left open and people do in fact 'meet in the hall.' Someone yells out through the tube and you have meetings that take place on the fly. It's an extremely valuable way to do distributed product development. It keeps people in sync and creates the feeling of one team rather than several teams.

"It [the Worm Hole] lets you feel much more tightly connected. However, it doesn't completely reduce the requirement for travel. You still have some need for being together and sharing meals and for very large meetings where you can't fit everyone into the lens of the camera. But it certainly reduces the frequency of trips."

Network of Partner Teams

The three-site hardware and systems software engineering group was only one team within a network of WorldMark development program partners. It comprised internal groups, external partners, and customers. First, there were the groups that were internal to NCR:

- The core development groups in California, South Carolina, and Illinois, which were supported by groups in two locations in India and one in China;
- The database development group in El Segundo, California;
- A communications software group in Lincroft, New Jersey;
- Marketing and administrative groups which were located at NCR's headquarters, Dayton, Ohio; and
- The manufacturing sites in South Carolina and Dublin, Ireland.

Then there were the outside partners:

- Intel in California and Oregon, which provided the basic processors;
- EMC in Massachusetts, which provided disk arrays that supply the system with memory capacity;
- Symbios Logic (which was a part of NCR when the project began but was sold to Hyundai while the product was being designed) in Wichita, Kansas, which also provided disk arrays;
- Informix in California and Oregon, which provided database systems for the new line of machines;
- Microsoft in Washington State, which provided operating system software support; and
- Many others including customers as well as additional suppliers.

Each of these partners had an internal team focused on WorldMark, with its own communications needs, as the system development group did. Sets of different partners then needed different mixes of connections depending on the work they were doing together (i.e., the need to exchange large files). The program as a whole needed to communicate across the many physical and organizational boundaries represented by the network of teams—the WorldMark teamnet.

Many Modes of Communication

Although few organizations today have such communications power available to them, the glimpse of the future that NCR offers carries important lessons for all virtual teams:

Use multiple media to offer many pathways for interactions and the development of relationships.

"You need to carefully blend in and utilize each tool for its intended purpose," Roberson says.

Voice

Because the telephone is so basic to work-at-a-distance, we scarcely reflect on how remarkable this voice-extension capacity is—and how it is continuing to evolve.

The WorldMark project made modifications to its voicemail system to ease the complexity that springs up among multiple players. "Voicemail had been around for a long time and it was a very good capability. But we decided that wasn't really good enough because we were doing team development and our teams were spread around the world," Roberson says. So they enhanced the addressing ability of their voicemail system to create one "virtual site." Usually, people have to dial different codes on the system for different buildings, cities, or countries and leave separate messages for each person. Instead, the enhanced system allowed people to send a single message to any subgroup of the team or, if appropriate, the entire team.

Video

Beyond voice connections that are basic to all virtual teams, video channels also can be interactive. For the WorldMark virtual team, the Worm Hole was not the only type of video link used to bind the disparate project group.

The team used "traditional" video conferences to link project partners, including the Indian and Chinese sites where high-quality telecommunications links are scarce. "This was one of the handy features of having AT&T as a part of us," Roberson observes. With AT&T's global communications infrastructure, the team minimized problems that most companies face in countries where telecommunication is still under development.

The WorldMark program also used desktop video conferencing—where a camera is mounted on the top of a person's computer monitor. Desktop video conferencing allows a small number of people to hear and see one another through small windows on their personal screens, as though they were gathering informally in a colleague's office.

Computer

In addition to numerous audio and video links, "the group also used more standard secured computer-based communication connections heavily," Roberson says. They used basic e-mail, available to everyone, with the capability of "embedding information"—the electronic equivalent of attaching multiple packages of any size to a letter. It was not atypical for an e-mail to contain ten embedded files, each containing 25 pages of documentation—marketing plans, service plans, detailed schedules, product specifications, spreadsheets, charts, even near-photographic-quality images. "The extent to which we embedded files probably took the state-of-the-art up a level," he says. The team also used intranet-based discussion groups to keep members updated on the project.

With all of this data and communication speeding around the world, the team naturally built its own knowledge center. It was linked to NCR's corporate information repository (data warehouse) that houses most of the relevant information for the company. The repository is available through conventional file access as well as through the company's internal Web-based intranets.

Face-to-Face

Like most virtual teams, WorldMark members also met face-to-face. Coming full circle, Roberson remarks, "We still met, of course. There were lots of meetings. You still need all-hands meetings. With all this wonderful technology and shared information, they still don't replace the need to get together with the whole team in a particular site and communicate with them on what's going on, on what the direction is, and on the importance of their contributions."

However, unlike many major virtual team projects, this one had no memorable kick-off event nor did the *entire* group ever come together face-to-face. "We didn't have to because we had greatly enhanced other forms of communication," Roberson says.

Communications Is a Process

As exciting and effective as the technology used on this project was, the WorldMark project had something else. It never would have hit the mark without a comprehensive planning and project management system that kept information flowing to the right people at the right time. Using what it calls its "Global Realization Process," the team was able to track and measure its progress monthly, weekly, and even daily. The process convened cross-functional team leaders for monthly meetings to evaluate where the project stood relative to the plan. They in turn reported their progress to a senior executive team that closely monitored development.

Even though this very large effort successfully combined all the elements of a virtual team—the people, the purpose, and the links—Roberson says that there was "still more to learn and more to do. The engineering side was done in a pretty spectacular fashion but we didn't have as complete a connection to all of the stakeholders as was needed. The people in sales groups particularly outside the United States could have been involved earlier and we had to mount a double-time effort with sales training. When you pull the schedule forward, you don't always pull everyone with you. It was a pretty massive development activity sited around the world and yet we accomplished it in record time."

Overall, links made the difference for the WorldMark virtual team—links of media, interactions, and relationships.

- WorldMark paid attention to its need for *physical connections* with face-to-face, audio, video, and computer media.
- The program managers set action items, articulated processes, and blazed new pathways for *boundary-crossing interactions* using its Global Realization Process.
- They laid the basis for strong *trusting relationships* that developed over time as people worked together.

Four Ages of Media

To operate effectively across boundaries, virtual teams become masters of media. They need to be media savvy in two very important ways. Virtual teams use:

1. "Process media" to run their own organizations, because the actual time the team spends creating, specifying, designing, and managing itself is largely informational work; and

2. "Product media" to deliver results, such as new products, decisions, reports, and plans.

Marshall McLuhan woke people up to the momentous impact of media on human experience with his 1964 book *Understanding Media*.[2] His memorable phrase, "the medium is the message," epitomizes his insights.

Imagine being asked to do something. Your interpretation of the request depends on whether the requester is your boss, subordinate, partner, or competitor. It also matters whether you receive the message in a face-to-face exchange, a handwritten private note, an e-mail, or a printed memo sent to everyone.

Many communications theorists separate the content of a message from its context. They point to the *metamessage*—the relationships, status, and interpretive cues that ride along with the literal symbols themselves. Scientist-philosopher Gregory Bateson called these bells and whistles the "command" part of the message.

McLuhan went a step further. He said that the transmission medium itself powerfully influences the total communications experience. That is, there is (a) the message; (b) the affect that it carries; and (c) the medium by which it travels—a meta-metamessage so to speak.

> *Perhaps the most basic message that a medium sends is whether it expects, allows, or makes possible a response. Virtual teams need to maximize their use of media that enable interaction.*

The Evolution of Communication

Just as a signature organization characterizes each great era of human civilization, so does a signature style of communication typify each era. Indeed, the rise of different media have usually been key features in differentiating the big break points in history.

Speaking, writing, and printing are the first three revolutions of human communication.

- Speech shaped the Nomadic Era and the formation of small groups and camps.
- Writing emerged in the Agricultural Age and made large-scale hierarchies possible.
- Printing spread specialized knowledge in the bureaucratic Industrial Age.
- In the information era of the Network Age, electronic media have shrunk the planet to McLuhan's famous epithet, the "global village."

Each organizational era—small group, hierarchy, bureaucracy, and network—brings its own capabilities that accumulate over time. Instead of new forms of organization wiping out the old, they incorporate them.[3] Thus, today's emerging network organization benefits from and includes the positive aspects of its organizational predecessors: the specialized functions of bureaucracies, the levels of hierarchies, and the coherence of small groups.

As the new impacts the old, it brings modern variations to recurring themes. The virtual team is a new form of small group made possible and necessary by new forms of communication. While we now have geographically distributed small groups, we still retain access to the variations spawned in each previous era. Command-and-control hierarchical teams such as military units and rule-based bureaucratic groups such as executive committees are still with us.

Successive waves of change may have reduced the globe's Nomadic Era populations to vanishingly small numbers. Yet the echo of communications in that age still reverberates distinctly in all human life at the turn of the 21st century.

Time has not diminished the importance of oral communications.

Writing, the second great leap in communication, incorporates the message contained in speech into a newly enduring, transportable medium. When written down, words persist and can move independently of the writer. Printing in turn incorporates writing while electronic media incorporate all previous media. The inventions of the new include the innovations of the old.

The One, the Many, and the Few

A fundamental distinction about a medium—whether it is one-way or two-way—shapes virtual team communication at every scale.

One-way media broadcast actions.
Two-way media enable interactions.

Because virtual teams must produce products and interact across distances fractured by delays in time, this distinction is crucial. One-way media are great for delivering products (and orders), but they do not enable the interaction (and good will) required for people to work in virtual teams.

Researchers typically characterize media by another distinction important to virtual teams: the ratio between the number of senders and the number of receivers. Broadcast television and the daily newspaper are one-to-many (1:M) media. Generally speaking, one entity produces the communication and many receive it. CNN cablecasts to you, but for all intents and purposes, you have no genuine ability to respond (even if a fax number or e-mail address appears on the screen). Contrast this with the telephone that people most frequently use in its one-to-one (1:1) form—just you and me. Interactive media also have their many-to-many (M:M) mode—such as the discussion and "schmooze" time at a conference or any large social event.

The range from one to many (1:M) senders and receivers is too broad to properly identify the media most important to small groups. Thus we add the category of the "few" modes of communication.

Virtual teams interact extensively in the world of the few.

Organizational life is replete with events that involve one sender with a few receivers—workshops, seminars, and briefings. Photocopiers, fax machines, and e-mail distribution lists are technologies that support one-to-few (1:F) communication.

Most important to the internal workings of all kinds of teams, but of particular import to virtual teams, are the media that connect a few senders with a few receivers (F:F). This is the communications venue for small group interaction, including the all-important medium of the face-to-face meeting. Virtual teams make frequent use of few-to-few conference calls, and of video conferencing, which is becoming more prevalent in both its room-based and desktop modes.

Online conversations, meetings, and conferences provide a new array of interactive digital media. They open abundant possibilities for communication among many senders and receivers—from groups of a few to vast numbers of participants.

Communication Media Through the Ages

Each type of media has certain common characteristics that influence effectiveness, cost, and accessibility of communications, the very stuff of virtual group life.

For virtual teams, the conditions for communicating across space and time boundaries are intimately involved with the nature of the technology they use and how interactive it is. Technology has moved the human world of small groups from the assumed state of collocation in place and time to the option of working together at a distance. This change is thousands of years in the making.

Virtual teams are the beneficiaries of this long evolution of communications technology. Media, once developed, do not go away. Although we no longer use stone tablets to communicate, it is generally true that we do not lose older forms of communication as we acquire newer ones.

Virtual teams today have access to a wide array of media and are able to choose among them for specific purposes.

It is also important to review past communications revolutions in order to put the digital difference of the 1990s in context—and to give direction to the onrushing future of virtual teams.

All the communication that people and organizations send and receive can be placed somewhere in the Communications Media Palette (Figure 4.1). Make your own matrix from the media available to you, or add in other media you know about.

Oral Media

Oral media encompass a full range of one-to-few-to-many channels that are both one-way and two-way. Speeches, workshops, seminars, and

Figure 4.1 Communications Media Palette

Nomadic Agricultural Industrial Information →

	Oral	Written	Printed	Analog Electronic	Digital Electronic
1-Way Active					
I : M	Speech Conference Briefing	Tablet Proclamation Manuscript	Book Film Newspaper Magazine	Broadcast TV Broadcast radio Videotape Audio cassette	Online broadcast Internet video Internet radio Online publication Digital packaging (disk, CD)
I : F	Workshop	Graffiti	Newsletter Memo	Photocopy Fax	E-mail list
2-Way Interactive					
I : I	Dialogue	Letter	Greeting card	Telephone Mobile Ham radio	E-mail File transfer Internet phone chat
F : F	FTF meeting	Flip chart		Voice mail Audio conference Video conference	Online meeting Online conference Intranet WWW Internet
M : M	Social event				

briefings are all one-way, sender-based; conversations, meetings, and social events are all two-way and interactive.

To speak to someone else without the aid of technology, both sender and receiver need to be in the same place (collocated) at the same time (synchronous). Consequently, a speaker can only reach as many people as the voice will carry to.

The physics of sound carries voice through the air. The receiver's capacity to hear and comprehend speech rule reception. Given the requirement of shared space and time, speaking offers a medium with no appreciable delay between sender and receiver.

People retain what they hear only in the private places of their individual memories, not in the communication medium that links them. Unlike e-mail, for example, which records all communication, speech evaporates. Reconstructing a remembered conversation has caused more than one argument. People interpret conversations privately, separate from the medium itself. In short, real-time oral communication has little inherent storage, recall, modification, or reprocessing features. Continuity persists through an oral tradition passed from memory to memory.

Written Media

The development of written languages, both ideographic and alphabetic, co-evolved (very roughly speaking) with the agricultural economy and the rise of hierarchical organization. (Egyptian hieroglyphics and calendars developed 5000 years ago setting the stage for the Early Dynasty period and the rise of the first great cities.) Writing offered message-senders options—from inscriptions on stone that have lasted for ages, to painstakingly produced and reproduced manuscripts, to the remarkably flexible medium of paper documents.

Writing represented a profound break with the limitations of the spoken word. Senders and receivers were no longer required to be in the same place at the same time; they could be in different places (distributed) at different times (asynchronous). The number of people reached by writing, while in principle virtually unlimited, was in fact quite small. The costs of production and transportation together with the literacy required for individual use capped the number of possible writers and readers.

Slower and more cumbersome than speaking, written interaction occurs when people exchange letters and notes. Delivery depends on the transport technology, which in the Agricultural Era included domesticated animals, wheeled vehicles, and boats as well as fleet-footedness. An individual's capacity to read governs the speed of reception. These characteristics add up to a general delay between sender and receiver, dependent mainly on the distance between them.

Writing freed communication from the constraints of space and time because of its most important quality in memory terms, the ability to be stored. Suddenly, human beings had a way to capture communications and make messages explicit, public, and permanent. While a great way to store ideas, writing-on-paper is still a limited medium in terms of its ability to help people recall or modify communications. (Witness how much time you spend rifling through files and piles looking for a particular piece of paper.)

Printed Media

Historians always cite the invention of the printing press and the production of the Gutenberg Bible in 1456 as key early developments of the Industrial Era. Printing is primarily a one-way medium, whereby single senders can reach great audiences of receivers through proclamations, books, and other printed materials. Newspapers, magazines, and newsletters are examples of print media in which a few senders (publishers, writers, and advertisers) reach large audiences of generally passive readers. Monographs or limited run publications offer some small-scale options, but until the advent of desktop publishing, the cost of production had been so relatively high that printed media have had limited value for interactive communication.

Like writing by hand, printing breaks the bonds of space and time. Unlike writing, print reproduction is comparatively easy; the time and cost differences between a print run of 1000 and 10,000 are marginal. Very large numbers of people are reachable through printed media.

Print production, however, is much more complicated and slower than writing. It involves not only the time required for writing, but also the time of transferring writing to the print mechanism and the time of printing the product itself. Speed of delivery is again dependent on the transport technology, which greatly increased in the Industrial machine era. Speed of reception, however, remains constrained by the speed of reading. These factors create what is usually a substantial delay between sending and receiving, rendering print media almost useless for sustained interaction.

Like writing, printing provides storage integral to the medium. Its recall, however, is still limited to remembering the location of the information and then physically combing through material to find it. Modification is, if anything, more difficult in printing than in writing.

Electronic Media of the Information Age

The paradigm-shattering concepts of relativity and quantum mechanics that announced the end of Newtonian absolutes and the coming of a new age of science are now almost a century old. Most of us point to the mid-20th century as the practical point of transition from the Industrial to the post-Industrial era. At the end of the century, it is a mainstream idea that humanity is now going through fundamental transformations in technology, culture, economics, and social organization.

We are now deep enough into the Information Era to begin to recognize major stages within this overall period of time. In 1964, McLuhan described the media of our time as "electric," remarkable by the almost instantaneous nature of communications based on principles of electromagnetism. Writing in 1995, Nicholas Negroponte, director of MIT's Media Lab, drew a fundamental distinction between *Being Digital*[4] and *being analog*. From Negroponte's point of view, analog TV shares more with analog books than it does with computer-based digital media.

In the analog world, "atoms" deliver information. We move molecules in the air, ship paper around, or modulate the structure of electromagnetic waves. In the digital world, "bits" deliver information. Bits are pure information, representations of on-off switches. They deconstruct the analog world into ephemeral strings of binary relationships and reconstruct them wherever. An analog book deteriorates over time, but a digital book is potentially timeless. An analog book occupies physical space, while a digital one occupies none the eye can see.

This very big difference between atoms and bits has a profound impact on virtual teams and their future development. Accordingly, we separate electronic media into analog and digital eras.

Analog Electronic Media

Broadcast TV and radio, videotapes, audio cassettes, and the like are all one-way analog electronic media. They allow senders to reach groups of receivers at virtually any scale, from local to global. Analog electronic reproduction extends to print through media such as photocopies and fax.

The extraordinary innovation of the telephone has made possible a new species of interactivity. It is the most important addition to the human repertoire of one-to-one communication since the evolution of speech. Audio teleconferencing and voicemail are group-oriented analog media. The same is true of "traditional" video conferencing and its offspring, video mail and desktop video conferencing.

People often remark on the distributed, aspatial nature of electronic media. However, this feature does not distinguish them from earlier non-oral forms. Senders and receivers of writing and print can be just as far apart as the people who communicate via electronic media.

In terms of time, however, there is an enormous difference. Electronic media completely fracture the constraints of time, offering synchronous or asynchronous connections, or even both together, such as recording a broadcast for replay. These media extend to virtually unlimited scales, reaching billions of people at the same time, such as during Olympic broadcasts.

Electronic communication effectively travels at the speed of light, a distribution speed that has no parallel in nonelectronic media. For production and reception, however, analog speed slams into real-time barriers. An hour's worth of information broadcast or viewed on a tape still takes an hour to meaningfully view (fast forward aside). How quickly people can speak and listen limit the speed of the telephone connection. This is the real-time limitation of the analog world.

In memory terms, analog electronic media offer little in the way of fundamentally new capabilities. Like writing and printing, electronic media can store communications, but provide limited support for recall and modification without additional digital capability.

Digital Electronic Media

ENIAC, the first electronic computer, was unofficially turned on at the end of World War II in early 1945 to help with some last minute calculations for the first atomic bomb. Thus, the birth of the digital era is linked with the nuclear explosions in August of that year that irrevocably sundered human time into "before" and "after."

Despite their dramatic entrance, computers stayed in the background for the next quarter-century, generally supporting the centralized, routine bureaucratic needs of the Industrial Era, fueling the rise of IBM to the pinnacle of global companies. Computers then shrank from mainframes to minis, led by then new companies like Digital Equipment Corporation. But it was not until the computer-on-a-chip escaped from the labs in the mid-1970s that the digital revolution began to flower and directly affect everyday working life. It gave rise to the now-ubiquitous personal computer (PC) and companies like Apple ("computers for the rest of us").

Somewhat simultaneous with the rise of the PC was the development of computer networks, initially created to spread out use of the incredibly expensive mainframes through time-sharing systems. These trends converged in the 1980s, heralded by the Macintosh, a PC with built-in networking. Networks are now the central computing paradigm. They link computers of every size and capacity, from massively parallel supercomputers to mainframes, minis, workstations, desktop PCs, portables, palmtops, and chips embedded in all manner of appliances. More than one company has used the slogan that Sun Microsystems has made famous: "The network is the computer."

The total computing facility available to society consists of both the computing capacity of individual devices and the network connections among them, what some call "the matrix."[5] This combination has given rise to "computer-mediated communications" in the parlance of early researchers in the field of *digital media*.

Like their analog counterparts, electronic digital media offer an array of one-way options, although many are only now being deployed as we approach the millennium, such as digital TV. Internet audio and video provide both one-way and two-way capabilities, although temporary limits of bandwidth are slowing growth. The transmission of graphics, audio, and video requires bandwidth that is vastly greater than that needed for transmitting simple (ASCII) text like e-mail, which is almost instantly replicable and can reach millions or a few. Digital content can also be packaged in CDs, for example, and documents can be posted for retrieval from online databases. The new media also offer something else: interactivity.

> *The digital media really shine in interactivity, exploding the limits to human organization and allowing a vast expansion in virtual group capability and variety.*

The options of one person communicating with another, of a few communicating with a few, or of many communicating with many others flow almost seamlessly from one digital variety to the next. E-mail ranks with the telephone and face-to-face dialogue as a powerful personal medium. Digital technology also allows the point-to-point exchange of files (including digitized print documents) and even replicates the telephone system through Internet phone. Small groups have a growing list of digital media available that allow a few people to communicate with a few others—from synchronous online chat and electronic meetings to asynchronous computer conferencing and topical discussions.

The scale of interactivity continues to expand beyond shared databases like Lotus Notes, where participants are both senders and receivers, to intranets, the World Wide Web, and the Internet as a whole. Digital media and especially the ubiquitous Internet represent an historically unparalleled expansion of interactive capability. Just what is different?

Digital Is Different

As with every media developed since writing, digital media support communications across space. Like analog electronic media, digital communication may be synchronous or asynchronous, and is effectively unlimited in terms of the numbers of people it can reach.

All computer-based media take full advantage of the speed of light. This is especially true at the nanoscale of the chips themselves, a level of functioning imperceptible to our natural senses. Production and reception speeds are not limited to real-time. They may vary enormously according to the type of data being prepared and communicated. A database instantly produces information at processor speeds while you spend real time typing an e-mail note.

The big difference in computer-based media, what makes it so effective for interaction, lies in its vastly increased memory capabilities. This pertains not simply in storage, which all post-oral media share, but also in memory's other aspects. Recall is integral to digital media. One can peruse vast quantities of information in moments, needles picked instantly out of the proverbial haystacks of data. Modification is unlimited; it is easier (and incomparably faster) to turn bits on and off than it is to retype a page. (Take a stroll down memory lane to compare the act of editing a document in a word processor with retyping pages on a typewriter.)

Reprocessing is unique to digital media.

No medium other than the computer-based one can reprocess its own stored information. Computer-based media can compress, split apart, and recombine information in infinite varieties. The medium itself makes possible computer enhanced images, data compression, packet-switching, language translation, content filtering, and morphing, to name just a few capabilities.

It is not just what you can do with the bits that is so exciting, but what you can do with the content itself. Of special interest are the digital connections that can link concepts, data, pictures, diagrams, and all manner of media. We have barely scratched the surface of the cognitive capabilities that digital media offer virtual teams and the organizational networks they undergird.

The digital medium is the ultimately flexible one. It can take on the shape and contour of any of the others, from a highly centralized mass medium to a completely decentralized interactive one. Most remarkably, it can be all forms at once, available to match the right medium to the right need. For a virtual team that is working anywhere at any time, digital technology dramatically expands its communication bandwidth—professionally, organizationally, educationally, intellectually, emotionally, and socially.

Media Characteristics by Age

Media vary by key characteristics. As we have seen, each great era of communication carries a common set of advantages and constraints summarized in Figure 4.2 along these dimensions:

- *Interaction.* How far apart people are in space and time and how many people the medium can reach influence interaction, the back and forth, or lack of it, of communication.
- *Speed.* The pace of message production, the speed of its transmission, and the rate of its reception govern the swiftness of communication.
- *Memory.* The ability to hold and use a message depends upon its storage, its ease of recall, its difficulty in modification, and its reprocessing capability.

Multiply Communication Options

Virtual teams need to know what their options are across the range of one-way and two-way media. Particularly, they need to distinguish between the communication required for their work internally and the media they need to communicate their work externally.

Figure 4.2   Media Similarities and Differences

	Nomadic	Agricultural	Industrial	Information	
	Oral	**Written**	**Printed**	**Analog Electronic**	**Digital Electronic**
Interaction					
Space	Collocated	Distributed	Distributed	Distributed	Distributed
Time	Synchronous	Async	Async	Sync/Async	Sync/Async
Size	Small	Small	Mass	Unlimited	Unlimited
Speed					
Produce	Speaking	Writing	Write and Print	Real-time	Variable
Deliver	Sound	Transport	Transport	Electronic	Electronic
Receive	Hearing	Reading	Reading	Real-time	Variable
Delay	None	Some	Lots	None	None
Memory					
Store	None	Integral	Integral	Integral	Integral
Recall	None	Limited	Limited	Limited	Integral
Modify	None	Limited	Limited	Limited	Unlimited
Reprocess	Separate	Separate	Separate	Separate	Integral

Virtual teams need both one-way "product media" as well as two-way "process media."

"I don't know what kind of maturity it takes to realize that you can't rely on just one medium to establish communication," says Bernie DeKoven, author of *Connected Executives*.[6] DeKoven believes that each medium is appropriate to a different kind of message.

"E-mail is the middle step between fax and phone," he says. "It's still informal but I am as accountable as if I'd written something down. I can misspell, and even be a little incomplete but whatever I've said can come back to me, and can be redistributed. E-mail has a certain viability but it's not as concrete as a fax. When I send a fax, there's a sense of permanence and formality that comes along just because of the medium of paper. It's a less tenuous form of communication than e-mail. It allows you to check, 'Is this what you wanted? Is this what you meant?' For you to sign off on it means we've achieved an understanding. E-mail is a longer term, less formal exchange—I say something and expect something back, then you add and so on. The point is to use each of the media for what they do the best."

One example of a long-lived virtual team that came to rely on a variety of media is at Hewlett-Packard (HP).

"What it really comes down to is massive communication," says Larry Banks, HP's Technical Education manager in its Medical Products Group. Banks is referring to the magic required to make virtual teams work. Like many others, he believes that "you can't beat collocation" and that the trick comes in making people "feel as if they are collocated."

Banks spent five years working as Research and Development section manager at HP on a project that crossed all the boundaries—organization, discipline, distance, time, and culture. Its purpose is to develop a product that makes it easier for HP's virtual teams to work together. The worldwide distributed product information management system (PIM System) that this geographically and organizationally separated project team works on allows HP engineers to manage schematics and mechanical drawings from design through manufacturing. Not all the engineers who work on a project sit next door to each other. Nor are all of HP's manufacturing sites in the same place. Thus, the PIM System project is key enabling technology for the company's virtual teams in the future.

Like the WorldMark program, the PIM System project is a good example of how many different kinds of links a virtual team needs to use:

- E-mail, which is continuous;
- Telephone conference calls, also frequent;
- Voicemail, which became pervasive and indispensable in the company during the life of the project;
- Face-to-face meetings (which always include an "element of celebration—food and fun," Banks says) that they hold three times

a year alternating locations between HP's headquarters in Palo Alto, California, and one of the other seven lead sites that are involved (including Germany);

- Real-time video conferencing, which allows people to sit at their desks and display what is on their computer screens to one another regardless of where they are; and
- An internal World Wide Web site.

While not every organization has the technology capacity of HP, most organizations, even cash-strapped nonprofits, have access to telephones, fax, and usually some e-mail. To work at a distance, over time, and across organizations, virtual teams must link copiously and variously.

Communication Pulled Apart

While communication has been critical to group life since the beginning of human time, today connective technologies are exploding exponentially. They enable small task-oriented groups to perform in extraordinary new ways.

"Links" is a short word for communication, the wonderful, vital term riding the global wave sweeping in the Information Age. Key elements of the future are wrapped up in multiple meanings of the word communication that pull in different directions.

- Communication means, first of all, a medium. If you hear people talking about the "Communication Industry," most likely they mean the collection of businesses that provide the technology and channels of communication—from hardware manufacturers to phone and cable companies to the broadcast industry. A "Communication Engineer" understands and applies the physics of communication. This is the meaning of the word we use in the Communication Media Palette.
- Communication also means interactions. A school of communication prepares you to be the intermediary between media and people in broadcast radio or television, or as a newspaper or

magazine journalist, editor, or publisher. More generally, every interaction "communicates a message" in the course of specific transactions.

- Finally, communication is a process of developing relationships. If people say they are "not communicating," most likely they do not mean that they have a technical transmission or message delivery problem. Rather, they refer to difficulty in their relationship. Psychologists recognize communication as the means of forming and maintaining relationships, involving deep issues of trust, reciprocity, and intimacy.

Communication has quite a diverse set of dictionary definitions in current usage. The word works, however, because there is an essential interdependency among all the different meanings.

Flowing from Concrete to Abstract

Like purpose, the concept of links spans a range of abstraction. At one end are concrete physical media like wires and telephones and even conversational airspace; at the other end are the often difficult-to-grasp relationships between people. While purpose flows from abstract vision to concrete results (see Chapter 3), communication links flow in the opposite direction. They move from the concrete tangibility of connections to the abstract intangibility of human bonds. Between physical connections and human relationships lie interactions, the moment-by-moment, blow-by-blow stuff of daily social life (Figure 4.3).

Links are physical media *that enable* interactions *that spawn and maintain* relationships.

Media provide the communication channels, the means of interaction. Channels exist quite separately from people or what they want to communicate. As technologies, they are passive and only offer the potential for communication, not the act itself.

Figure 4.3   Communication Links

Media	Interactions	Relationships
Technology		People
Medium	Process	Patterns
Physical	Observable	Experience
Data	Behavior	Learning
Nervous system	Thinking	Knowledge

Interactions, on the other hand, are all about process. To communicate is to interact; to interact is to communicate. Interactions are not separate from the people involved and how they interpret experience. They are also behaviors that generate public information for observers. Researchers study interactions to understand the dynamics of groups and teams.

Relationships represent the cumulative effects of interactions, however few or brief. They are the patterns that simplify the complexity of human interactions, the learning and emotions retained from the intensity of direct experience and fed back into future interactions. Over time, relationships develop among people in a group because of their interactions with one another, eventually enabling them to become a team.

Mapping Relationships

One day early in the life of Apple Computer's new global software engineering organization, vice president Steve Teicher called a small meeting of senior staff. He wanted to map the web of relationships that already enmeshed the fledgling group. He went to the white board, which covered a 15-foot-long wall, and started to draw circles with names as people kept up a steady stream of suggestions. By the time he was done, dozens of interconnected circles ringed the board.

Relationships among all the people and organizations involved with a virtual team can add up to a staggeringly large number of possible permutations. They comprise every combination of the people in the group plus innumerable linkages outside the team—the whole web of the team. Mapping detailed relationships inside and outside even a small group of people can become frightfully complex, a task that is marvelously managed by the science of social network analysis.

A relationship never belongs to one single person or another but to both people together. Although relationships exist between people, they do not occupy any physical space. They grow over time, may span years of inactivity, and yet may fracture in a moment. Our relationships are at once the most durable, the most fragile—and the most rewarding— parts of our lives. Relationships among the members are the bonds that enable virtual teams to do their work across boundaries.

Massive linking begins to suggest that cognitive metaphors point to real opportunities for smarter organizations.

- Media provide the nervous system for the virtual team.
- Interactions along these pathways by members of the virtual team constitute the team's "thinking," the shared communication that is audible and public to the group.
- Relationships are patterns of interaction where a virtual team accumulates its long-term learning. The fabric of relationships and shared knowledge are the internal stabilizing forces of virtual team life. As one virtual team leader says, "It's one part technology and nine parts context." He is stressing the importance of the environment of relationships in which the virtual work takes place.

With relationships, we move from the domain of technology to the realm of people.

CHAPTER 5

TEAMING WITH PEOPLE

The Paradoxes of Participation

We live in a 270-year-old house near Boston, Massachusetts, and we still get milk delivered weekly in glass bottles. These are all anachronisms: the pre-Revolutionary house, the milk route, and the glass bottles. The rest of the developed world goes to the supermarket to get its milk and juice in paper cartons and plastic containers. Most likely the packaging comes from Tetra Pak, the half-century-old firm that delivered the first milk cartons in 1952 and now provides its products in 117 countries.[1]

In Fall 1994, the company's subsidiary Tetra Pak Converting Technologies AB took a bold leap into the future. They eliminated internal functions altogether and reorganized around client project teams. No more line managers and no more staff.

Tetra Pak Converting Technologies

Tetra Pak Converting Technologies (CT) is a 115-person independently incorporated company within the 18,000-person agglomeration of 50 factories and product companies that is Tetra Pak. One of the biggest packaging manufacturers in the world, Tetra Pak produces 75 billion packages a year. It is the largest of the four branches of Tetra Laval, the Swedish foodstuffs giant formed in 1993, a 35,000-person behemoth.[2]

Tetra Pak sells two types of products, machines that fill the packages and the packaging materials themselves. CT, working at the juncture of machines and materials, is a production engineering house that creates new converting equipment and helps factories develop new processes to reduce costs.

As globalization spread in the 1970s and 1980s, Tetra Pak began to open factories around the world. This trend accelerated in the early 1990s. Suddenly, the bulk of CT's clients were no longer in Lund, Sweden. The company had to adapt—quickly.

Transformation at Light Speed

When Sture Karlsson arrived as managing director of CT in August 1994, he began his tenure by talking to the company's clients about their needs and expectations. Initially confused by what he was doing, employees quickly came around. He engaged the whole company in the client discovery process and the subsequent discussion of how best to organize. They learned that clients wanted more projects done faster and better.

At the same time, Karlsson talked individually with CT's employees. He found out that CT's major business process was not engineering or research and development or any of its other functions, but rather its client projects.

The take-off point for the new organization came at a two-day management meeting in October 1994, just two months after Karlsson arrived. "We started to think, 'If the project is our focus, why do we have a line function?' That was the turnaround point," Karlsson says.

Most experienced managers say change takes time, inevitably more than anticipated. But not always. "It can go faster than you expect. I was surprised that it went so fast with such commitment all the way. A great portion of it was the way that we communicated it," Karlsson remembers.

Within a day of their decision to change, management sat down with the unions to say they needed to reorganize but "didn't know how it would end up." Two days later, they took the same message to the whole company thereby establishing a pattern of continuous communication. Karlsson's one-on-one meetings continued. Small groups held open discussions in "square meetings" (referring to the architecture of the office centers), and monthly companywide meetings took place (for those first two turbulent months, they took place weekly).

Three questions immediately arose when they asked themselves what the risks were in moving to a purely project structure:

1. How could there be a stable place to have salaries set and make social contacts?
2. How could they be certain that they had a process for long-term competency development in place?
3. How would connections happen between the teams and across the company to share knowledge and manage common processes and resources?

To address the issues related to people, they developed a mentor system. Employees would choose a member of the management group to act as their direct link to the senior team, to help develop their individual competence development plans, and to act as the key figure in setting their salaries.[3]

Networks address core competency and infrastructure needs. Each network has a sponsor from the management team and a competency leader who is a non-management specialist in the area.

Role Redesign

As the discussion about the new organization continued, decision making gradually expanded beyond the nine-member management group. The workshop finalizing the vision and organizational structure involved 27 people, a quarter of the company.

People experience organizational change as a change in their roles. At CT, they have continued and strengthened some roles, such as that of project manager. Other roles such as line managers have gone away. There also are altogether new roles including network and competence support.

Change hit the management group particularly hard. Their functional roles had been eliminated. Karlsson allayed their fears by making it clear from the start that "roles are changing, not people." The management group as a whole was charged with redefining its place in the new structure. Together they defined the key new roles for themselves:

- Mentors to 10–15 employees each;
- General project support;
- Project sponsorship and steering committee membership;
- Network support;
- Factory (client) contact; and
- Factory and product company contact.

The CT management team takes responsibility for a variety of initiatives: internal networking, the overall direction of the company, the total results of project activities, and the external network of clients, suppliers, and competency sources, as well as the relationship with the larger Tetra Pak organization.

Project teams are another key element in the new structure. Multilevel by design, each team includes a management team sponsor. The sponsor is also part of the team's steering committee along with the client and other stakeholders and advisors. The project manager personally commits through the project life time, the traditional role having been weighted with considerably greater responsibilities for success.

Specialists staff the teams. They also belong to and may share leadership in at least one competence network. Two sorts of competency networks serve systemwide needs:[4]

- Core technical competencies related to the specific requirements of the business (such as printing); and
- Role (project managers' network, secretaries' network), infrastructure (communications), and enterprisewide functions (such as quality and the environment).

At CT, the purpose of a network is "to maintain and further develop work skills and competence." Everyone is expected to keep one another informed about projects, literature, courses, exhibitions, study visits, suppliers, and other external contacts that bear on the network's specialty. A Competence Support leader convenes, coordinates, recruits, and speaks for the network. Activities are small in scale so as not to burden the projects, and time is allocated for support work that everyone recognizes as requiring time unconnected to projects.

Project Routines, one of CT's networks, shows how a role-based network contributes to quality, improved processes, and cost reductions. In this example, as project managers discussed how to run projects within CT, they began to build a common file of best practices, guidelines, and standards for projects. The network is also responsible for the competency development of current and future project managers.

One of the notable savings of the new organization is a dramatic shrinkage in capacity buffers. These are the underutilized resources that most functional organizations accept, such as people sitting around not productively occupied. Four factors provide this flexibility for CT:

1. Management team project support that coordinates new projects, responsibilities, and resources;
2. Guidelines allowing teams to borrow staff from each other for up to a week, and longer with the agreement of project support;
3. A common source of work methods and "project routines" that enables specialists to easily move between teams; and
4. Networks where people discuss resource allocations of present and anticipated work with an openness "nonexistent in the old organization."

Going "First Class"

Communication in all of its many meanings is the key to CT's success. First and foremost has been communication of the change process itself. Once Karlsson and his management team glimpsed the new structure, they switched to the new model while working out details along the way with the people directly affected. The process both demands and generates trust. "If you are insecure about where it will go, then the process itself must be very secure," says Karlsson.

Once the change was underway, face-to-face interactions increased dramatically in one-on-one meetings, small group discussions, and all-company gatherings. Even as the direction and plan took form, CT recognized the need to expand the number and type of communications channels to support the project teams and competency networks.

In mid-1995, less than a year after the change process had begun, CT started to use First Class.® This groupware system offers sophisticated e-mail, conferencing, and newsgroup bulletin board services. Each network and many projects established their own conversations. Online conversations are open to the whole company. Monthly team reports are also open to everyone.

The impact of questions being posted in the morning and answered from 10 directions by evening generated early excitement and underscores the business value of the online exchanges. "It is a tool for us to make network thinking obvious," Karlsson remarked. "It's the type of experience that makes you feel this is the right thing to do. It helped me show my boss what we were doing. We let him in [to the discussions] just before Christmas, and he suddenly said, 'Now I understand.'"

CT did not rely solely on computer-based communications tools that soon included the Internet and the use of internal Web sites. They put together a plan to utilize other channels as well, including large meetings, an internal news-sheet/magazine (featuring summaries of management meetings), cascading information through the reporting structure, gatherings, literature, and individual interviews. Their media plan indicates whether the communication is one-way or two-way and how often it occurs (for example, monthly large meeting, weekly magazine, continuous newsgroups; Figure 5.1).

The most important changes, however, have been in perspective and behavior. In virtual teams and the networks that they connect to, the overall communications pattern shifts from delivery to access. Lena Bengtsson, responsible for CT's communications competency network, says, "We have transformed the whole system of information flow and have tried to change the habits of our colleagues. Instead of being fed with information we encourage people to be curious and seek out information. Now our responsibility is to see that the information required is available."[5]

Figure 5.1   Tetra Pak Media Plan

Media	Type	Interaction	Frequency
Large meeting			Monthly
Internal magazine			Weekly
Cascade information			Weekly or biweekly
News groups			Continuous
Gatherings			Monthly
Literature			As needed
Individual interviews			Variable

In virtual teams and networks, each act of sending information is an act of leadership that requires making assessments of need and appropriateness. An open information system puts new demands on people, which boil down to two at CT:

1. "Do I need to make sure that this information reaches specific receivers? If [so], the information must be clearly addressed.
2. "Is this information that I want to be available to people when they need it? If [so], then I must store the information correctly."

Extensive communications support offers great benefits for regular and routine information exchanges, for providing background and related information, tracking plans, and the like. In Karlsson's view, old-fashioned face-to-face meetings are still best for "gaining commitment and problem solving."

The Stress of Being Me and We

CT has had a unique and instructive way of balancing the inevitable strain between individual and group needs. At CT, the company addresses all three aspects of the life of the individual in the new boundary spanning world: independent members, shared leadership, and integrated levels.

While it is not easy to be a member or leader of a team, it is even more difficult to play these roles in a virtual team deep in the flux of change. All the self-doubting questions that any team member asks—"What am I doing here? Do they need me? Am I included? Who thinks they are a leader here? How aggressive do I need to be? Will I measure up?"—can be even more exaggerated when the group lacks daily face-to-face contact.

Doubts, concerns, perceived problems, and boredom mingle with excitement, opportunities, caring, satisfaction, and even exhilaration. To be part of a team is to continuously work a dynamic tension deep in the heart of being human.

People must simultaneously be "me," an independent individual, and "we," an interdependent part of groups.

Each of us grapples with an inevitable and continuous tension between the need to *differentiate*—to enhance our individuality—and the need to *integrate*—to bond in groups.

Complements Not Opposites

Individuality is necessary for cooperation. A paradox. An apparently contradictory assertion that may be true.

Too often the individual and the group are posted at opposing ends of a contradiction, each vying for primacy in a win-lose contest. We characterize entire cultures as individualistic (for example, the United States) or group oriented (for example, Japan).

In reality, *me* and *we* are complements, not opposites. This is the key to resolving the paradoxes of participation.

Virtual teams are high-connectivity/low-maintenance organizations.

To a significant degree, virtual teams are self-managing. To be successful in virtual groups, people must have much more independence and decision-making capability than people typically do in bureaucracies. People who form teams that cross boundaries need to know more, decide more, do more. This is made possible by clear purpose and personal commitments together with open, accessible, comprehensive information environments. These in turn link to the ongoing conversation that is the team's process.

Sture Karlsson puts it this way, "People must know more about the vision and purpose when they cannot lean on the side of the organizational box they belong to."

It gets more complicated if you are simultaneously a leader of teams of subordinates and a member of teams of peers and bosses. "Me" is me personally, but also me representing "my team." "We" is the family feeling of "me and my reports," but it is also the language of "me and my peers" with the boss. How can people be both "me" and "we?"

The Janus View

To see *me* and *we* across the boundaries of a virtual team, everyone needs the ability to adopt a "Janus view." It is a personal and fundamental virtual skill.

Janus is the Roman god of beginnings and endings, the guardian of doorways. The god of portals has two faces, one that looks in and the other that looks out.

> *Janus views life from the boundary—looking inward to the group itself and looking outward to other people and other groups.*

The CEO has a natural Janus view. The top-level leader sits on the organization's boundary and is skilled at balancing views of internal needs and capabilities with external assessments and strategies. Internally, the organization as a whole appears as a web of relationships, while externally a web of relationships enmeshes the organization itself. Not only at-the-top leaders, but leaders at every level sit on boundaries. Simultaneously they peer "up and down" and "in and out."

From the Janus view, people are *holons*. Holon means *whole* ("hol-") and *part* ("-on"). People are both wholes and parts. People are parts of groups and may stand for the whole group as leaders.

Arthur Koestler originally coined the word holon.[6] It concisely expresses the idea that everything (like atoms, cells, solar systems, cars, and people) is simultaneously a *whole* in and of itself and a *part* within larger systems.

Usually called "hierarchy" by scientists, the holon is a central principle of general systems theory. It is the idea that life and the universe and everything in between structures itself in levels, "subsystems comprising systems within suprasystems." Mathematicians talk about "sets-of-sets." Nobel Laureate Herbert Simon called hierarchy the "architecture of complexity."[7]

Simple word, complex idea. We use the holon (hierarchy) idea every time we use money, outline a report, store a file, find a reference, or check an organization chart. We use the holon idea when we "go up a level" to a higher authority, broader scope, and more abstract view. We also use it when we "go down a level" to more detail, narrower scope, and more concrete views.

In virtual teams, people operate as holons in three ways, as:

1. *Members,* the parts, whether people or groups of people;
2. *Leaders,* the connective tissue between the parts and the whole; and

3. *Levels,* the successive wholes that make up complex networks, the recursive idea embedded in the holon.

Strange and new a word as it is for most people, holon can stand for organizations, small groups, and individuals. It is logical (if a bit strange) to say "a team is a holon composed of individual holons that are part of a larger organizational holon."

Stripped to its mathematical essence and used in the context of technology, a holon is a "node." People and virtual teams are nodes in networks. A node may be simple—one person—or it may unfold into a whole universe—America Online is one node on the Internet. A team is a node in a larger organization, and it comprises member nodes linked into a network. This ability to map organizational terminology to technology is a powerful benefit of using the virtual team model (see Chapter 7, "Virtual Place").

Members, leaders, and levels are three transformations that resolve the me/we paradox. They turn flesh-and-blood huggable people into intangible hard-to-grasp virtual teams.

1. The transformation of the autonomous individual into a member of a team;
2. The leadership transformation of individual members into the group as a whole; and
3. The transformation of a group of individuals into a "group individual," a new level, a team.

Independent Members: "Who Is Involved?"

The first transformation rests on what seems to be an uncommon-sense idea: *People are not the only parts of groups.* Does this make sense? It seems so obvious, beyond question: People make up groups! Period. However, there are problems with the view that people are all there is to groups:

- It obscures the reality that the group is something more than the sum of its members. A virtual team is a unit—a coherent system itself—or "something more" that is separate from and in addition to its corporeal members.

- If people are the only parts of groups, then the ability to analyze and understand groups in a detailed way is limited by individual human psychology and the ability to peer inside people's heads.
- Finally, if only flesh-and-blood individuals can be members of small groups and teams, then there is no meaningful way to talk about groups that we perceive as organizational individuals, what anthropology might call "fictive individuals." The law formally recognizes corporations (which American English even refers to in the singular—"IBM said today that it would . . .") and nations ("France declares . . .") as "individuals."

Roles Relate People

People are not parts of groups the same way that hearts are part of people's bodies. Only in the extreme (for example, slavery) does a group own people body and soul. Lynda Popwell's experience at Eastman Chemical Company of finding herself on too many teams is not unusual. Most people are members of multiple groups. We all take part in a constantly changing personal pageant of many small groups simultaneously—family, community, friendship, and affinity groups as well as task-oriented work teams. In each group and team, we play different roles.

> *Like people, roles are integral parts of groups. People animate roles that belong to the group.*

The role is a basic social structure that mediates between an independent individual and expected behavior in a group. Roles naturally arise in small groups and are what sociologist Erving Goffman calls the basic "unit of socialization." In a small group, roles are informal, more "felt" than "visible." In larger organizations, however, roles tend to take on more concrete trappings through titles, written job descriptions, and personal contracts.

Although you cannot see them, you experience the importance of roles by talking about your part in a group: "What is my role?" or "That role's already filled," or "I can fill that role," or even, as you are leaving, saying, "There's no role for me."

> *Roles translate between me and we, between the bottomless complexity of individual people and the comparative simplicity of playing a part in a group.*

Roles are easier to see in their more formal manifestation as "positions." People usually diagram positions in relationship to other positions, for example, an organization chart where this person reports to that one. They often accompany them with written profiles—job descriptions. Positions clearly belong to the organization that sets them up and can just as easily take them away.

A position is either "open" or "filled." You receive "an offer" for a position that you "take" or "accept." An open position—a formal role—stands by itself as a sometimes gaping hole in an organization, an empty place in the structure. When a person steps into a position, a classic dynamic arises between the characteristics of the particular person and the legacy of expectations that the role conveys. Once populated anew, the role both shapes and is shaped by the person who occupies it (Figure 5.2).

Formal positions provide clues to the informal roles that people play in small groups that can be more elusive. People (particularly management) also carry their positions into the many teams they join. Sometimes this is appropriate; sometimes it is not. For positional teams, such as an executive management team—a *de facto* task group because of its place in the hierarchical structure—it is especially important for people to understand both their formal and informal roles.

In virtual teams with limited face-to-face interaction, roles rise in importance. Consider that in virtual teams:

- People typically play multiple roles, often many more than in conventional teams;

Figure 5.2   Roles Integrate "Me" and "We"

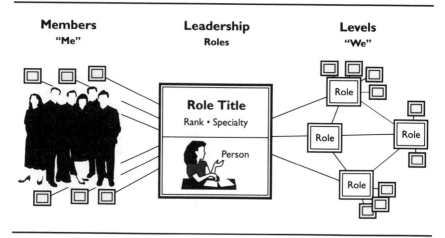

- Roles require greater clarification. Expectations need to be made more explicit than they are in collocated teams; and, at the same time,
- Role flexibility is essential. Because the process is dynamic, roles are constantly changing.

Independence Starts with Me

Respect for the individual is a core value of all the great team companies. At Motorola, for example, its management philosophy "begins with two key beliefs—respect for the dignity of the individual and uncompromising integrity in everything we do."[8] The trick is to develop greater cross-boundary capabilities (through clear purpose and constant communications) without diminishing—better yet, while enhancing—the independence of individuals and teams.

Enhance independence as you strengthen interdependence.

Independence is a quality that permeates every level of organization—from the personal level of people as members of teams, to teams as member parts of larger organizations, to the independence required of companies in alliances. That is, all groups need a minimal level of independence and decision-making in relationship to the larger system. Virtual teams need even more.

Independence can never be complete or absolute. Not for people, teams, companies, or nations. Independence is always a matter of degree along a range from "too little," to "sufficient," to "optimal," and finally, "too much."

When CSC Index, the Cambridge, Massachusetts, consulting firm reorganized after an unparalleled growth period, it drew specific attention to this tension. "The local offices in New York, Atlanta, Boston, Chicago, and San Francisco needed autonomy," recalls senior vice president Judi Rosen, "but we also needed strong ties among the people in specific practices who were spread out among the locations."

Because virtual teams need higher levels of *interdependence* in roles, they require correspondingly higher levels of relative *independence* and voluntary behavior in the individual members.

For virtual teams, entirely new roles have sprung to life to deliver productivity and provide cohesion, such as network support at Tetra Pak Converting Technologies. It is not just new roles. People must play old team roles in new ways. This is particularly true for the central role of leadership. There the struggle between "independent me and interdependent we" becomes part of the group persona.

Shared Leadership

One leader makes for a good sound bite, but it takes more than one to lead a successful virtual team.

Insofar as the sudden proliferation of virtual teams is in some ways a harking back to a simpler way of organizing, it is instructive to look at how the most original teams handled leadership. In forager societies, there are many informal leaders. Among the !Kung tribe in the Kalahari Desert in Botswana, a foraging society that has survived thousands of years in spite of tremendous threat, leaders influence but they do not force.

Traditional anthropology interpreted such systems as being without a head ("acephalous"). Then in the late 1960s, University of Minnesota anthropologists Virginia Hine and Luther Gerlach confirmed that this distributed leadership form is in reality many-headed ("polycephalous").[9] Herbalists, hunters, midwives, warriors, and other particularly skilled or knowledgeable people take the lead as circumstances require. To one frustrated researcher trying to identify a single local leader, a !Kung elder said, "Of course we have headmen! . . . In fact, we are all headmen . . . . Each one of us is headman over himself!"[10]

Virtual teams take a page from the !Kung book. As organizations that require much more leadership than conventional teams, when successful, they nevertheless have much lower overall coordination cost. This only works if everyone understands and assumes part of the expanded virtual leadership burden.

Grasping a Group

We each wear many hats, a typical metaphor for diverse roles. Even very small groups may have members with many overlapping roles and the number of possible roles is infinite. Decades of research on small groups and teams searching for general team roles have turned up this major insight: The only universal role observed in groups is leadership.

> *Virtual teams that are highly self-motivated and self-managed are leader-*ful *not leader-*less.

Leadership is pervasive in virtual teams. The leadership structure as a whole is an inclusive set of related roles of leaders and followers. Reuben Harris, chair of the Department of Systems Management at the Postgraduate Naval Academy, has identified six basic leadership roles that virtual teams require:

1. Coordinator
2. Designer
3. Disseminator
4. Tech-net manager
5. Socio-net manager
6. Executive champion

The transformation of a person into a group by way of a leadership role is a miracle of social construction. Leaders are convenient handles to help members and outside observers alike grasp groups.

When confronted with complex ideas, people have a habit of using one part of the idea to represent the whole.[11] "Wall Street" stands for the complexity of U.S. financial markets; the "Oval Office" stands for the presidency and Executive Branch of government.

The phrase, "I belong to Jim's group," shows one person representing a whole group, nowhere more obvious than in the role of the CEO. Here, a person stands for a corporate entity that may include thousands of people, "speaking for" the organization externally and "speaking to" the group internally.

The habit of simplifying complexity by grasping a prominent part translates into a presumption of single-pointed leadership. Cultures even build in this view. Such is the case at one major company that requires every project to have a single designated responsible individual or DRI.

While virtual teams may have single leaders, multiple leaders are the norm rather than the exception.[12] Virtual teams that deal with complex issues and problems invariably have shared leadership regardless of the titles they use for convenience.

Many authors of books on teams simply assume without discussion that a team needs a single leader. A few distinguish, as we do, between formal leadership (governance) which may be singular and the broader multiple leadership that always arises in a successful, healthy team. "In successful teams, leadership is shared," states Glenn Parker unequivocally.[13]

In the earliest teams, the camp teams, leadership was informal and distributed, based on influence rather than authority. We are in many ways returning to the organic structures of that era, albeit with a fantastic new capability to create nonterritorial spaces and share information.

Social and Task Leadership

Small groups typically have at least two kinds of leaders—social leaders and task leaders, a distinction first made in the 1950s:

- *Task leadership* is oriented to expertise, activities, and decisions required to accomplish results. The measure of task success is *productivity*. This leadership clearly is of central importance to virtual teams, since "the task rules" in this type of small group.
- *Social leadership* arises from the interactions that generate feelings of group identity, status, attractiveness, and personal satisfaction. The measure of social leadership success is group *cohesion*.

In a traditional hierarchy-bureaucracy, social leadership simplifies and formalizes as a place in the authority structure. Task leadership boils down to one core expertise. A typical role title reveals both the social and task aspects. Consider the vice president for Manufacturing:

- The vice president is a designation of social *rank*, a level in an authority structure—the hierarchy part of the title.
- Manufacturing is a label of task specialization, pointing to an area of expertise—the bureaucracy part of the title.

How do you convey rank online? New interactive media such as e-mail pose unforeseen problems to the existing authority structure. In work areas, for example, space displays importance (a closed office versus a cubicle), signs offer titles, and choice of attire differentiates employees from executives.

Rank—having it and using it—is a major challenge for virtual groups.

While this status-creation process exists in virtual teams (as it does in all teams), the role of rank is far from clear or easy to settle. Too much rank ossifies the team all too quickly. Rejecting it sometimes cuts the team off from necessary organizational connections. Throwing out hierarchy blindly also risks the loss of the crucial navigational and cognitive advantage of levels.[14]

A team will not coalesce or feel complete until it identifies a critical mass of appropriate people with the expertise required to accomplish its tasks. Teams acquire skills for their tasks as people perform their activated roles.

A new team often defines its expertise roles before it locates the members who populate them. This is in itself a step toward virtuality. Imagine a team that does not yet exist. It is most often the search for the "right people," those with needed expertise and experience, that leads to different locations and organizations—and the consequent formation of a virtual team.

> *While rank is confusing, specialization is booming in virtual teams. Your area of expertise most often defines your role in task-oriented virtual teams.*

"I can't think of any project that we do on our own. There is just too much to know and there are too many specialties in the built [architected and constructed] environment," says Gary Wheeler. Wheeler is a principal at Perkins & Will/Wheeler, the Chicago-based architectural, engineering, and interior design firm and a past president of the American Society of Interior Design. "Ninety-nine percent of what we do is not a stand-alone. We designed a sales office in New York that involved an engineering firm, a construction manager, a real estate consultant, an audiovisual consultant, a move coordinator, and technology and data consultants. The 'real' client was in Cupertino, California, and the user client was in New York but even they were in three different divisions. Our job was to make sure that the whole group was interacting and coordinated. People had to step forward and then step back when it wasn't their job."

Managing the challenges of virtual team life also brings the opportunity to involve the best minds and most experienced people, wherever in the world they may be. In time, great teams will become the norm as we climb the learning curve of distributed work.

Integrated Levels

Big organizations are made up of smaller organizations that are made up of small groups. Small groups tie together organizations from the front line to the executive suite and board room.

As the basic unit of organization, how big is a small group? How big is a group of small groups? Does being virtual make a difference in size?

Counting the Guests at the Virtual Table

The number of people on a team is one of those things that appears so obvious that it is easy to miss its significance. All teams, after all, have a size. Size refers to the number of people who are members. Size also strongly influences the internal communications burden and the number and variety of interactions and relationships that the team requires.

The size of a collocated team is rather immediately obvious and membership is usually clear. In virtual teams, size often can become fuzzy as membership swells and contracts as individual participation peaks and wanes. Virtual membership boundaries often have degrees of "centralness" or "bands of involvement"—for example, a core group, an extended team, and external partners (Figure 5.3).

> *Millions of years of experience indicate that some numbers recur. There seem to be two natural breakpoints in the size of small groups: 5 and 25.*

The number 5 sits at the approximate midpoint of a range for the size of a team. Researchers, popular writers, and experienced team leaders

Figure 5.3   Bands of Involvement

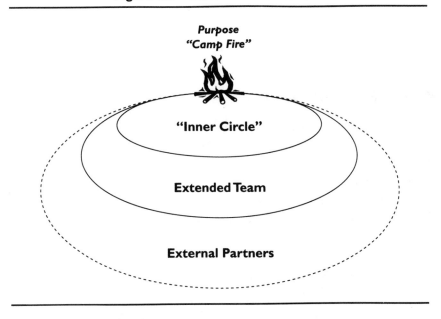

Purpose
"Camp Fire"

"Inner Circle"

Extended Team

External Partners

alike agree that the ideal team size ranges from 4 to 7 members. This is, not so coincidentally perhaps, the same size as a typical Stone Age family and not very different in size from many families today.

Is there a lower limit to team size? One debate among researchers is whether two people, technically known as a dyad, are enough to be considered a group? Three people, so some thinking goes, bring enough diversity to qualify as a small group: Three people offer multiple communication pathways and the possibility of subgroups and cliques.

This is not a question for us: two can team. If we look just at the roles that people play in groups, then even two people can play many roles with one another, with a great diversity of communication between them. As friends, lovers, spouses, parents, business partners, and even co-authors, we surely are a very small but very complex group.

Is there an upper limit on how big a team or small group can be? Different people suggest different numbers, but the general upper limit figures range from 15 to 25. Some writers offer a different sort of rule for measuring the extent of small, such as "the number where everyone knows everyone else," or whatever size can form a "functional unity." Teams of 25, however, typically are groups of small groups.

Teams Cluster into Teamnets

Teams do not naturally exist in isolation. For millions of years, teaming occurred in camps and groups of camps. Teams naturally belong to a camp (Figure 5.4). This remains true today, even if the camp is often unrecognized.

The nomadic family yoked together between 4 and 7 people as its basic socioeconomic unit, the same size as today's typical team. From

Figure 5.4   Early Evolution of Team Levels

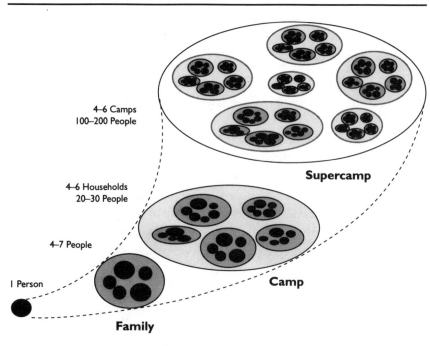

time immemorial, these small units naturally have congregated into larger associations. Camps, involving clusters of 4 to 6 families, appear to be as universal as the family itself. The Olduvai Gorge in Tanzania, for example, reveals base camps of 25 to 30 people as early as 1.7 million years ago at the very beginning of the Stone Age, the Lower Paleolithic era.

This primordial clustering has given rise to what researchers have called "the magic number 25,"[15] camps of 5 families averaging 5 members each. Twenty-five is also the number of people in most everyone's "persisting life-long network." These are the folks who are closest to you throughout your life—job changes, divorces, births, deaths, moves.

The size of the camp is comparable to the outside limit for a small working group. With more than 25 or 30 people, a comfortable meeting becomes difficult and starts to turn into a conference and people cease to be entirely familiar with one another.

At the next level, Nomadic Era camps invariably joined up in a "supercamp," a local network of 4 to 10 or so camps who together identified the foraging territory of a "local group."

These supercamps are comparable to a large group of 100 to 200 people, another natural cleavage point in modern organizations. W.L. Gore & Associates, the folks who brought Gore-Tex to the world, keep their plant size to a maximum of 150 to 200, which founder Wilbert ("Bill") Gore believed was the number at which human achievement peaked. Larger than that, he said, people start to get in one another's way.

When people call a group that is bigger than a handful or two of people a "team," they usually are referring to a "team of teams." This is a group that has a common set of cross-team goals and interdependent tasks—what we have dubbed a *teamnet,* a network of teams.[16] Understanding the appropriate internal team structure is an often overlooked design issue. People often make these sometimes contentious subgroup definition decisions too early, too make-it-or-break-it-confrontationally, or too unconsciously and off-handedly.

A good yardstick for team size is the "rule of about 5": Package work for small teams of 4 to 6 people. On average, 25 people will work in 5 5-person teams.

> *There cannot be one "right" size for teams. Team size depends first on the task at hand, and second on the unique constraints and opportunities of the situation.*

Generally, the more complex and diverse the task, the larger and more diverse the team needs to be—more expertise, more people. While more people bring more talent, they also bring along the need for more coordination that generates its own problem. Adding people helps performance up to a point. Then the law of diminishing returns sets in. Before long more people degrade performance.[17] After a limit, which seems to vary by task, more people may actually do less. Sound familiar?

Big, big, qualifier: Since these rules around size come from millions of years of experience with collocation, it is only a starting point for estimating the appropriate sizing and clustering for virtual teams.

Virtual teams can be successful only if people cooperatively manage the coordination involved in membership and leadership. With the skills and infrastructures in place to multiply and share leadership, we are seeing some teams explode the apparent limits on productive size. Virtual teams tend to have small active core groups and large memberships.

No Team Is an Island

Engaged distributed leadership leads to new levels of organization. New levels arise both from team integration and team differentiation (Figure 5.5).

A collection of individuals who begin interacting interdependently on a task over time can become a team. The identity of a new team becomes confirmed as people begin to use the words "we" and "our." Sometimes there is a moment when the team coalesces, a clear "click" audible to all participants. Sometimes, people are more aware that they became a team in hindsight.

As important as people are, the achievement of "teamness" is the creative act of a group, not an individual. Relationships persist *among* people not in them:

- A new level is born through integration: the team "pops" into existence separate from its members. This is the miracle of synergy in systems, the living result of something more than the sum of its parts.

Simple teams have two-level structures, but most teams, even small ones, develop three levels over the course of time. As a team begins to plan and perform joint tasks with diverse specialties, typically overlapping subgroups of a few people form so that they can pursue several strands of work concurrently.

- New levels are born through differentiation, when internal groupings form as the work unfolds. Each is itself a team microcosm with a need for clarity of purpose and communications.

For fast, flexible productive virtual teams, the work must shape clear internal organizational structure. Indeed, it is in their internal work design that the intelligence of the group is manifest. The process,

Figure 5.5   Virtual Team Levels Ruler

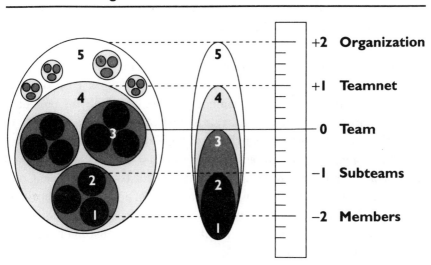

categories of work, and relationships shape the interactions and ongoing conversation that is the team "thinking out loud."

With each new level, new team roles and responsibilities emerge. A group with an identity itself becomes an "individual." The team acts and is perceived as a unit at the next level of organization. Indeed, teams that are really humming often become very inwardly focused, sometimes creating bonds that rival family ones in strength.

Warning: Team success can breed team insularity.

Management movements like quality and reengineering have created a new myth: *the team as hero.*[18] While this is a great recognition of the renewed importance of small groups, it also tends to invest the team with rampant, often competitive, isolationism. Independent teams without inter-team interdependence can fragment corporate structure.

We are in danger of moving from isolated bureaucrats sitting in specialized boxes to isolated teams of specialists.

The *team-alone* syndrome dominates many businesses as well as other organizations. Individual teams spring up as challenges arise that the existing hierarchy-bureaucracy cannot manage. Generally unconnected to one another, these teams are rarely part of a conscious strategy to grow the organization to meet the challenges of accelerating change.

Some companies are already working in 21st-century, virtual team style. For Tetra Pak Converting Technologies, Eastman Chemical Company, and Sun Microsystems, virtual teams are over time a key business strategy. They offer competitive advantage for meeting challenges of speed, cost-effectiveness, and quality in a global, customer-focused, rapidly changing economy.

CHAPTER 6

IT'S ALL IN THE DOING

Virtual Team Life as a Process

A team is first and foremost a process: It has a beginning, a middle, and almost always an end. No team springs to life full-blown and none lives forever. Words such as conception, gestation, birth, childhood, adolescence, adulthood, mid-life crisis, and old age all apply to a team's life. Powerful results accrue when virtual teams consciously work their way through a lifecycle process.

Growing a Strategy

ServiceCo (a pseudonym) is one of the largest facilities management companies in the world. With operations in 35 countries and revenues of $3 billion, ServiceCo manages computer operations, building and grounds maintenance, and other support services for corporations, educational institutions, and healthcare facilities. Since its 1985 start as a small player in the burgeoning European facilities management industry, it has grown to be an international corporation with 100,000 employees that dominates many of the markets in which it competes.

By its nature, a facilities management company is the archetype of a networked organization comprising multiple virtual teams. ServiceCo's teams operate inside its clients' organizations. They perform tasks for which the client has no special competency: Sydney's Rocks Bank's expertise is in banking, Massachusetts' Art Academy's strength is education, and New Jersey's Eldercare Medical Center's specialty is healthcare. Thus all of these institutions outsource the operation of their facilities management to ServiceCo.

What ServiceCo offers is functional expertise. It employs a network of experts who are available to provide technical support to accounts as needed in software development, purchasing, landscaping, and security, to name a few. District managers who know the facilities management business oversee and mentor account managers working directly with clients. In addition, the company packages its expertise as standards, procedures, and computer systems that tie together their far-flung operations.

Most ServiceCo employees work at the clients' sites and have far more contact with client personnel than they do with the ServiceCo organization. Account managers spend most of their time managing client relationships and the local labor force. Most district managers and even regional vice presidents spend almost all of their time on specific account-related matters. As a result, most ServiceCo personnel feel as much a part of their clients' organizations as they do ServiceCo's.

To maintain connections among their decentralized personnel, ServiceCo builds communication and shared purpose through what it calls its "Strategic Process." ServiceCo Group, the parent organization, follows a tradition of holding annual structured planning efforts that look out over the next three years. Such strategic planning helps the subsidiaries keep their action plans aligned with their individual long-term strategies. In addition about once every five years, each subsidiary goes through its own major Strategic Process, during which it rebuilds its strategy from the bottom up through virtual teams. By involving people at all levels of the organization, the company gains widespread support and commitment that make for effective and energetic implementation.

Rethinking Health

In 1990, ServiceCo Group expanded its U.S. market with a series of acquisitions. They included a Texas company with clients in business, education, and healthcare (primarily large hospitals) and a Florida-based company whose business came largely from nursing and retirement homes. (In 1993, the two companies were merged and renamed ServiceCo U.S.A.) The extremely complex and dynamic U.S. healthcare market caused the international firm to re-examine its basic assumptions. Among the issues it had to consider were shrinking use of hospital facilities, intense cost pressures, and rapid realignment and consolidation in the industry. In particular, it needed to rethink how it provided service to acute care hospitals and long-term care institutions.

Sophisticated hospital and nursing home systems were demanding their facilities management providers take more of a virtual team approach. They wanted to coordinate services among widely dispersed facilities and integrate across service functions. To sharpen its approach, ServiceCo U.S.A. embarked on its first strategic process in the healthcare divisions. ServiceCo U.S.A.'s CEO Sol Kramer recruited Whitewood Management Consultancy, an English firm with U.S. operations that had worked with the parent group in Europe. On this project, the consultants acted as process, methodology, and quality leaders, filling in where analytic skills were weak.

ServiceCo U.S.A.'s New Market Strategy Group (NMSG), comprising its healthcare division presidents and the senior corporate staff, became the official body for the Strategic Process. They set three overall goals:

1. To develop a breakthrough "ambition" and strategy for the healthcare sector that would be broadly shared and supported throughout the company;
2. To develop planning and analytical skills among people in the healthcare division; and
3. To create relationships across the geographic and functional divisions that would facilitate the greater integration clients were demanding.

Initially, the NMSG chartered eight virtual teams. Three were to look at the needs and buying practices of different client segments. Three were to consider the clients' customers, that is, patients. Another was to examine healthcare "payers" (health maintenance organizations, insurance companies, and the like). The final team was to develop market statistics for the other teams. Each team drew members from across geographic and functional divisions of the company.

Getting Started

To facilitate a quick start, the teams received considerable structure along with their charters. Members were designated either as "core" (expected to devote 20 percent of their time to the team) or "backup" (expected to spend 5 to 10 percent of their time on the team). Each team had shared leadership. A designated leader was responsible for managing the overall plan and timely delivery of products; an "issues leader" was responsible for the quality of the informational products that were their parts of the strategic plan (for example, reports, data, analyses, decisions). To supplement its charter, each team received detailed "issue analyses." These analyses provided clear categories for a preliminary work breakdown of their problem areas. The teams were expected to meet face-to-face at least once to develop a preliminary work plan that they would present at the kick-off meeting one month later.

For some teams, the first meeting produced the required plans. Others took longer, struggling to understand their charters and formulate an approach to their work. Whether fast or slow in creating their first draft plans, all were ready for the kickoff at The Art Academy, a ServiceCo client located in a Boston suburb. The decision to hold the kickoff at a client site was deliberate: It symbolized the importance of client focus in developing strategy.

The goal of the all-day meeting, according to Eric Rogers of Whitewood, was "to clarify the purpose of the overall project and to enable the teams to see how their individual charters and plans contributed to it." The session mixed presentations by senior management and the consultants with workshops where the various teams presented their plans and got feedback.

That evening, the group moved from the school's newly constructed theater (that provided numerous break-out rooms for the day's workshops) to the oldest building on campus: the school's president's house. With its high ceilings and rich dark wood beams and paneling, the evening's venue and agenda were very different from that of the day. The goal of the evening session was to build relationships. Because the divisions had operated so independently—in "separate chimneys" as they call them at ServiceCo—many people on the teams had never met one another. To speed up the socialization process and ensure a memorable event, the facilitators orchestrated a murder mystery theater over dinner.

By the end of the 10-hour day, team members knew that the company was making a serious commitment to defining its strategy. The individual teams had refined their plans and had started to define their end products.

The Pattern of Teamwork

After the kickoff meeting, teamwork began in earnest. Team members worked independently on agreed-upon tasks, coming together to review one another's work and make decisions, then splitting up to do more concentrated work. Between face-to-face meetings, the groups held periodic conference calls to check on progress. Face-to-face meetings continued to be very important particularly since a major goal of the project was to establish relationships among normally dispersed managers. Most meetings concluded with dinner for the whole team. Frequently, two teams met at the same time and would merge at the social interludes, giving a chance for even broader interactions.

Early on many of the teams appointed someone to be the official "nudge." The original thought was that this person would be responsible for calling other team members to make sure they were on schedule between meetings. As it turned out, the nudge's most important role was to document the results of the team meetings. By capturing what transpired, the nudges helped solidify the evolving understanding of each team's purpose and its method of expression in the products it produced.

As is the case with most teams, the ServiceCo teams soon began to build their own task-oriented in-talk: They called a school system a MISO (standing for Multi-Institution School Organization). One team gave nicknames to its members while another adopted a cheer and a mascot.

The teams relied heavily on voicemail that proved to be quite convenient for people who spend their lives traveling to accounts. E-mail was new to the company when the project got underway. "To be sure someone read their e-mail, you had to send them a voicemail," recalls Ethel Berlin, a ServiceCo market development manager and team leader. With time, however, as the teams worked toward completing their products, e-mail became indispensable as a way to move drafts around among team members.

Phasing In Work

The original plan called for the teams to disband after accomplishing their original work in Phase 1 and reform in new teams. When the time came to transition to Phase 2, aimed at understanding the competition, the NMSG decided to stick with the existing groups. The teams had bonded so well despite their virtuality that the NMSG chartered eight new overlapping teams with one representative from each original team to study competitors.

The Competitor Teams never met face-to-face: They used conference calls to coordinate data collection, then shared their results via e-mail and overnight delivery. The competitor-team members then went back to their original teams where they compared competitor competencies with market requirements. The original teams (with some changes) then continued into the third phase where each group developed and proposed strategy options.

At the end of each phase, the team leaders and issue leaders presented their results to one another and to the NMSG. Phase 3 marked the end of the "public" part of the Strategic Process. By then, more than 75 people from across the subsidiary had participated in the virtual teams.

The NMSG now retired to the "private" decision-making part of the strategy development process. With the team recommendations and options now synthesized by Whitewood, the NMSG developed its own unified options and scenarios for the future. Over two months and seven full-day meetings, the senior group made choices and crafted its own vision and strategy for the healthcare market based on the work done by the teams. The strategy had major implications for structure and re-sources. With the will-to-act and unity of purpose generated by the pro-cess, there was little resistance to making the required changes.

At lower levels in the organization, the process created confidence in the future and in the organizational outcomes. Participants could all point to elements of the strategy that their team had contributed and could claim ownership for various aspects of the final product. The long-term implications of the project for the smooth operations of the com-pany were perhaps its greatest benefit. "Getting to know people from different parts of the organization was tremendous," says Marvin Krieger, Human Relations manager for one of the healthcare divisions. "I had never really worked closely with my counterparts in other divi-sions before. This project will make it much easier to integrate across di-visions in the future."

The Team Pulse and the Life Cycle

Virtual teams are living systems not machines. Made up of people with interdependent roles and a web of relationships aligned through shared purpose, everything about them is organic. As *living* systems, they are not biological organisms but rather social organisms,[1] which have both a pulse and a life cycle.

The proper metaphor—living system or machine—is critical to the understanding of virtual teams. It is hard enough to get face-to-face teams to "happen," to "jell" over time. It is doubly hard for virtual teams.

Teams grow. They take time to develop—and virtual teams tend to take even longer.

The Rhythm of Aggregation and Dispersion

A team's life cycle has its own rhythm oscillating between interludes when members come together and when they go apart. This tempo obtains through the long-term patterns and peak moments of key gatherings, the overall life cycle, and the hour-by-hour cycles of a team's daily life.

We still can hear the echoes of the earliest groups in human history in organizations today. While archaeologists cannot excavate social organization in the same way that they can unearth shards of pottery, they can infer a lot about it. By matching artifacts with direct observation of foraging societies that survive today such as the !Kung of the Kalahari Desert in Botswana, we have a reasonable facsimile of the "organizing process" for the first teams.

There was a pulse to the ancient life of nomads: groups of families came together and then went apart. Foragers had to follow the rhythm of the seasons dictated by their sources of food. Even today, !Kung households move to the same beat that literally "goes with the flow." Access to water moves the !Kung through seasonal cycles that cause groups of families to diverge and converge. The !Kung beat holds for the

!Kung Seasonal Cycle

From December to March during the hot, rainy summer season of Bara, !Kung families disperse to the maximum as food and water are widely available. As April and May, the cooler and dryer fall season of Tobe, approaches, the families begin to gather in camps around the larger water holes. From June to August, the cool and dry winter of Gum, several camps cluster around one of the permanent water holes, which define the locality. They remain there through September and October, the warmer but still dry early spring of Gaa. As the hot late spring of Huma brings showers in October and November, families quickly disperse into temporary camps. There, they take advantage of water caught in the tree hollows of the mongongo groves. As summer comes and water is plentiful once more, families scatter over the territory as the cycle begins anew.

way most people work—coming together and going apart. People work alone and then join up in a group. We do what we do best independently and then work with others to expand our capabilities. The basic social rhythm of human beings has not really changed in two million years.

The !Kung's major camp gatherings are akin to business off-sites. These are special times and places for convening teams to literally "pull things together," to resolve conflicts and decide future actions. They are also times of intense social interaction. Some managers regard the community-building aspects of such meetings as so important that they insist on them in spite of tight budgets. As we inaugurate the age of virtual teams, such meetings become all the more important. Most of the people whom we interviewed for this book stressed the importance of face-to-face interaction to solidify virtual teams.

Face-to-face time is increasingly precious, a scarce resource in limited supply.

When the !Kung families come together, they suddenly find themselves living in a very different environment with a greatly increased local population and numerous channels of interaction. Their camps are alive with feasts and dancing, partying and ceremonies. Suddenly there are many hands to make light work. People hunt together and build common storage facilities, share resources and information, trade goods, and exchange tools. Perhaps most important, the camps are incubators for new families, where people make matches and find mates.

Camps of 25 and supercamps of 100 to 200 serve broad human needs to associate with other people. Multifamily (the analogy in business is multifunctional and multiteam) camps arise from exchanges, interdependent relations, and repaid reciprocity. There is an ancient and natural tension between the family (the team) and larger social groups (organization). Even so, the cooperative act of sharing across kin (organizational) lines is a critical, necessary step in the development of human societies (networks) of all sorts.

Cooperation and Concentration

The "together/apart" rhythm vibrates deep in all sorts of human groups. People congregate then separate not only over seasons but in the course of a day as well. Think about your day with some of your time spent alone and some time spent with others. Time-lapse videos in Steelcase-sponsored research show a remarkable pulse to team life. Collocated teams of people come together for a time, then separate to do individual work—a together/apart fluctuation that replays many times over the course of the day.

An old-line office furniture designer and manufacturer that now sees itself in the "work performance" business, Steelcase characterizes this working rhythm as "times of cooperation (together) punctuated with periods of concentration (apart)." Among the many office systems they have designed to facilitate this natural pattern of interaction is their Personal Harbor and Commons product.[2] People work privately in their own Harbors and gather to collaborate in the Commons.

"The design absolutely facilitates communication," says Loree Goffigon, director of Work/Place Strategies in the Los Angeles office of Gensler, the architecture and design firm. The consulting group that Goffigon works in uses four of the Steelcase systems. "The doors on our individual Harbors are closed only 5 percent of the time. We have lots of space to pin things up around the activity tables in the center. We can hear what one another is saying and we call back and forth while we're on the phone. We're trying this out for our clients who want to collapse the literal and figurative time between the transference of ideas and information."

> *Virtual teams have a harder time getting started and holding together than collocated teams. Thus, they need to be much more intentional about creating face-to-face meetings that nourish the natural rhythms of team life.*

Establishing the life pulse is not hocus-pocus. Life is sparked by the sequence of activities that people undertake together and continue apart. It lives in how we choose to start things, whom we invite to participate, what agendas we create, what plans we make, which tasks we implement, when we reach milestones, and how we bring closure. A team unfolds through its unique life cycle.

Forming, Storming, and All That "–orming"

Team life is a process. Most organizational researchers and authors acknowledge and underscore this small group truth. Popular and academic studies alike agree on the general outlines of team process. Many books on teams use the model (or a variation on it) of the "stages of small group development" developed by Tuckman in the 1960s:

- Forming;
- Storming;
- Norming;
- Performing; and
- Adjourning (often omitted).

This nearly two-decade-old model retains its original freshness because it accords with the experience of countless facilitators and team leaders who have used it as a guideline. Among them is Apple CEO Gil Amelio who used this model for his turnaround of National Semiconductor.[3]

"Stressed S"

The Tuckman Model also has a powerful theoretical basis. It is a social variation on a growth model that applies to everything from astronomy to biology to marketing. The "S" curve (in mathematics it is the logistic growth curve) is so common that Ludwig von Bertalanffy, the "father" of general systems theory, offered it as the original proof that there are mathematical principles and patterns that hold across diverse sciences.[4]

Virtually all successful team process models follow this universal cycle of life whether consciously or unconsciously.

When applied to a team, the "S" curve has some interesting ripples. The Tuckman Model points to an important, overlooked feature of the life cycle, times of natural turbulence and potential conflict—stress points. By anticipating the likely stress points, a new, still-forming team gains a powerful advantage. Team members can use these natural points of commotion to give their process the energetic lift it needs—or they can be thrown off-balance by conflicts that seem to come out of nowhere. While not all conflict is predictable, some of it is.

There are two major points in a team process where stress is predictable—near the team's beginning and not long before its end. The Tuckman Model incorporates the first stress point (storming and norming), while the second most famous process model—the quality sequence of "plan-do-check-act"—points to the turbulent testing phase as "checking."

While the quality cycle gives no hint of the anticipated tumult in the early stages, the Tuckman Model misses the difficulties that often arise in the later stages of a team's life cycle (test).

The "Stressed S" is a generic process model (Figure 6.1), which we labeled elsewhere as the general stages of network development—Start-up, Launch, Perform, Test, and Deliver Phases.[5] In *The TeamNet Factor* we offer several methodologies for using this process model, from simple to comprehensive, along with supporting tools for complex implementations.[6]

The Pulse and the "S"

Team life is dynamic, managing the tensions of stability and change while moving forward through the life cycle. That same root dynamism lives in each of us, the conflicting pulls of being both "me" and "we." A significant way that this plays out in team life is in the pattern of aggregation (we) and dispersion (me).

Virtual teams must be especially conscious of their dynamics. Behavioral clues are spread out not only in space but usually over longer timeframes than they are with comparable collocated teams. Virtual teams

Figure 6.1   "Stressed S" Team Process

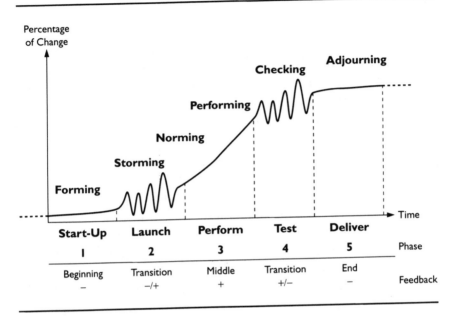

need to design for this supercharged eventuality. Smart virtual teams, like ServiceCo's Strategic Process teams, develop methods that antici- pate a life cycle and accommodate its predictable moments of stress.

From the perspective of feedback, it is apparent why these stress points occur in the life cycle. Peter Senge, who brought systems dy- namics and organizational learning into the center of contemporary management thinking,[7] describes the complementary feedback loops found in every process as "slowing" and "growing" actions:

- *Slowing* is the dampening, stabilizing, conserving tendency that keeps change in check; and
- *Growing* is the building-on-itself, accumulating tendency that expands change.[8]

Change and growth are processes of going from one level of func- tioning to another. Stability must be disrupted in order for change to occur. Then things restabilize. In virtual teams, feedback loops go from slowing to growing and back to slowing over the course of a life cycle.

- *Phase 1: Start-Up (Slowing).* During the initial forming stage, slowing prevails as the idea for a team and its initial formation struggle to arise against resistance. The team's initiators generate interest, gather information, and explore ideas. This phase may have an excruciatingly long fuzzy beginning that no one clearly recalls or it may have a breathtakingly brief, well-remembered "aha" initiation. Either way, change of any kind must struggle against the status quo.
- *Phase 2: Launch (Transition).* As a critical mass of people and purposes comes together, the team's ensuing storm begins to brew. Before the team is really ready to perform, it must sharpen its vague purpose, establish leadership, make plans, find resources, obtain commitments, and acknowledge norms. This is the first transition poised between the slowing loops of Phase 1 and the growing loops of Phase 3. Launch is the "make or break" phase. During this period, the team establishes the code for its life cycle and sows its seeds of success and failure. Some teams never get out of this phase and there are no guarantees here. It always takes painfully longer than anyone thinks that it will, and for virtual teams it can take even longer still.
- *Phase 3: Perform (Growing).* Most teams would much prefer to start right here in the growth phase. Perform is where the team does the bulk of the work. This is where results accumulate and where the team makes progress toward the goals set in the launch phase. People meet and overcome obstacles. At its best, life is good and seemingly will go on forever. But growing cannot go on indefinitely without being checked and reshaped by countervailing slowing actions.
- *Phase 4: Test (Transition).* This is the challenge phase, where the team must review results, finalize features, and limit resources. Meanwhile, time is running out and customers are demanding the goods. This is the second transition phase, with the

process now going from growing to slowing. Early participatory planning, customer involvement, regular reviews, and milestones can turn this phase into a triumph. All too often, this late-in-the-game trial is unexpected. Some teams never make it through this phase.

- *Phase 5: Deliver (Slowing).* Deliver is the endgame, the adjourning phase. The team delivers results, provides support, wraps up details, and ceremonializes its endings. The dominant tendency here is slowing as the team seeks to stabilize at a new level after it establishes change or a development cycle completes. It may be the end of one lifetime and the beginning of another and its duration may be brief or long.

Creating Time Together

"I believe that you clearly expedite [team processes] by spending more time on the front end and getting consensus," says Eastman Chemical Company CEO Earnest Deavenport. "You shorten the implementation cycle as opposed to the opposite when differences and resistance come out in implementation."

The moral for virtual teams who want to design their together/apart pulse is simple—and widely held by experienced team leaders and experts alike:

Invest in beginnings.

You will recoup time spent in the first two phases many times over in later phases. Mistakes, mistrust, unexpressed viewpoints, and unresolved conflicts all too easily introduce themselves and become part of operating norms. Lack of clarity around goals, tasks, and leadership hobbles the team in the performance phase. Failure to establish criteria and measures for results ensures a rocky ride during the inevitable testing phase regardless of whether the team is collocated or virtual.

Anticipation is the recommended prescription.

"Optimally a team gets together at the beginning of its process," says Curt Crosby who coordinates the virtual team effort across Sun Microsystem's largest division, Sun Microsystems Computer Company (see Chapter 7). "It's the old forming, storming, and so on. It doesn't always happen like that because at a company like Sun it's too easy to start these grassroots teams without meeting. If I can catch it fast enough, I do recommend that they get together at the beginning, mid-point, and at end of the life of the team." Crosby's advice:

- Invest face-to-face time for the start-up and launch phases.
- Reserve time for a meeting to assess and review the team's work before it completes.
- Punctuate your process with breakpoints and milestones where the team converges and realigns its work.
- And celebrate at the end. Even when the team has not been an unqualified task success, you may accrue valuable social capital (see Chapter 9).

Teams that take the time for ceremonious closure provide their members with valuable information about how they worked together as a group, Crosby observes. "If everyone is informally assessed by their peers," he says, "they have that experience to take to their next team assignment. This potentially shortens the storming cycle that might occur based on an individual's perceived weakness and willingness to overcome it."

Aggregation is a major challenge for virtual teams. Some have no face-to-face time at all. If the team cannot all meet together, perhaps one person can act as liaison-in-chief, as Buckman Laboratories' then-CEO Bob Buckman (see Chapter 2) did. When the company was launching its online knowledge network of virtual teams, Buckman traveled around the world, playing the role of the "circuit rider" carrying news from outpost to outpost.

Most of the virtual teams that we interviewed use telephone conference calls to provide some means of synchronous meeting and many relied on video conferences. The people at Buckman Labs found, as have many other companies, that a very active online conversation can be fast-paced enough to seem almost real-time. Buckman's early chat rooms allowed people who had never met (and might never meet) to have "screen" conversations where people talked about their families and hobbies. The major advantage of these sessions is that they quickly build a modicum of trust and usually cause affection to develop among the participants as they glimpse one another's private lives.

Sun Microsystems uses integrated digital environments that bring together features of chat with shared computer screens and the telephone. Intel is pioneering the uses of desktop video conferencing for virtual teams. Technologies that work well for small face-to-face groups and capitalize on the peculiar strength of the digital era are driving the explosive growth of teams and team capabilities. Intranets combine all the digital media into "digital" campgrounds. These "virtual water coolers"[9]—reminiscent of the !Kung gathering around Kalahari water holes—offer entirely new options for shaping meaningful aggregation in virtual teams while supporting their dispersion.

Forming Goals

Purpose sustains and initiates process. It is the source of life for all teams, the inner fire that gives them their vitality. Here, virtual teams face two particular challenges that differ from those of collocated teams: First, purpose "costs more" both in terms of the length of time it takes to develop and in the literal cost of bringing together distantly situated people. Second, purpose plays a relatively more important role as a legitimizing source of authority than it does when the boss is watching. On the other hand, purpose well set can also be a source of economic benefit: Coordination costs lower when empowered people align around goals.

For resource-lean and information-rich virtual teams, the more the design of the organization mirrors the work plan, the better. As the team carries out goals, the organization re-forms to address new goals and the next pieces of the work. Thus, the organization is constantly re-forming, organically adapting to the dynamic unfolding of the work.

Forging Cooperative Goals

Virtual team success or failure begins with the relationships among people and goals. Nearly a half-century of empirical research demonstrates the power of cooperative goals in determining team success.

Social psychologist Morton Deutsch was the first to use goal interdependencies as a way to predict how well people would work together. He asked whether people saw their goals as cooperative, independent, or competitive relative to one another.

- *Cooperation* occurs when people have compatible goals. When you succeed, I succeed. Confidence and trust are the expectations in behavior. Cooperation generates positive feelings of family and community as people share and integrate information.
- *Independence* results from the belief that goals are not related. Your success or failure has no bearing on mine. I do not expect any support or hindrance from you. Aspirations are personal and relationships with others are impersonal. We all do our own thing and have no need to share information.
- *Competition* follows from incompatible goals and the belief that if you win, I lose. Your success diminishes mine. I not only expect no help but I anticipate hostility and prepare accordingly. To prevail in competition rather than be integrated in cooperation, people hoard information and use it as a source of power.

Dean Tjosvold, Professor of Business Administration at Simon Fraser University in Canada, has been at the forefront of team researchers bringing a wealth of learning from hundreds of studies into real-world practice.[10] He reports that myriad studies document this simple fact:

Cooperative goals motivate team members.

When goals are compatible, people strive to succeed and the work required becomes meaningful. Performing tasks and reaching goals cooperatively bring the added benefits of helping others, feeling good, and storing goodwill for the future. Cooperation spurs the sharing of information and increases the insights available for planning, problem solving, and executing. People who work cooperatively are confident of success and believe that others want them to do well. They have more fun which translates into more positive feelings about work. Most importantly, a wide range of studies over all age groups shows that cooperation results in higher productivity than competition or independent work. This is particularly true for problem-solving and related tasks.[11]

Researchers' conclusions about competition surprise no one. It does not motivate people to share information, plan together, or find the best path for producing results. Competitors do not expect others to help or encourage them. Competition motivates people who believe they have superior ability and are likely to win, but it demoralizes people who have (or believe they have) lesser abilities and experiences. Competition also can motivate teams where tasks are simple and information needs are low, providing most of the people believe they have a chance of winning. However, since sharing information is the lifeblood of a virtual team, competition within hinders or scuttles success.

Designing Tasks

Whether intentionally designed or not, tasks and rewards will always generate either cooperation, independence, or competition.

- Group tasks promote *cooperation* that is strengthened by joint rewards. When they are in the mode of cooperation, people assume that everything is fair and that they will be rewarded accordingly. They pool their talents, offering and using individual skills and competencies as needed by the tasks. People appreciate creative conflict as a tool for finding the best answer. "We had some pretty heated arguments," says Bill Crowley who led an award-winning SunTeam in the company's SunExpress division (see Chapter 1).

- Unrelated tasks that are separately rewarded encourage *independence*. The measure of individual success is explicit external criteria such as quotas or sales targets. People use their abilities to further their own goals. They avoid conflict, regarding it as a distraction from separate pursuits.
- Codependent[12] tasks—separate pieces of work that *require* a winner and a loser—create the environment for *competition*. Such systems need rules to regulate the games. People use their abilities against others. They avoid conflict entirely or deliberately escalate it to gain personal advantage.

Most work situations involve a mixture of these motives, which are always complex. Tasks that are set up interdependently require cooperation. At the same time, people compete for attention, praise, promotions, and raises while also taking pride in their individual accomplishments.

Typically, people encourage cooperation within a team to better compete with outside groups. One familiar archetype of this behavior is the great sports team—for example, the Boston Celtics basketball team in the 1970s whose internal teamwork was legendary and enabled them to win championship after championship. Such us-against-them behavior is considerably more tricky in work organizations. Many a successful team that bonds into a tight family also excludes and competes with outsiders. Unfortunately, outsiders to the team may still be insiders in the organization. A company with many teams ultimately wants all of them to cooperate for the good of the enterprise.

However cooperation fares inside the corporation, can we still safely assume competition takes over at the enterprise boundary? For hundreds of years, the simple rule has been to cooperate internally and compete externally. Even this maxim has been challenged. Countless alliances explode across corporate boundaries. Networks tie companies closer to vendors and customers. Competitors cooperate on a range of issues from common interests supporting an industry to saving money together to joint research. In explaining a May 1996 meeting that he and Apple CEO Gil Amelio had attended with Microsoft Chairman Bill Gates, Apple's COO Marco Landi said, "We live in a world where your toughest competitor must be your best partner."[13]

Interdependence: From Cooperation to Competition

Interdependence is not only a feature of cooperation, but of competition as well. One person alone does not a competition make. Contestants are *codependent*, requiring a loser in order to be a winner. Competitive conflicts develop from differences in people's personalities, motivations, fears, perceptions of the facts, opinions, interests, and how much power they wield. At the extreme, large groups of highly organized combatants fight each other to the death—which is war.

In some situations, companies deliberately set up teams so that people hold individually or departmentally conflicting goals. Digital Equipment Corporation, under its founder Ken Olsen, was famous for intentionally setting up competition among product teams under the belief that such rivalry would improve results. This is now regarded as a very costly approach to creativity, and in the early 1990s, Digital began to change and bring teams together to resolve conflicts. For example, five highly competitive Digital teams suddenly were required to work together and resolve their differences before a major trade show. Although the intense month-long collocated process was at times contentious, it ultimately was highly successful and customers heard a unified voice about one product.

The path to cooperative payoff often leads through the thickets of competitive disagreements. Indeed, this is where the truly excellent team shines—in moments when they meet their greatest internal challenges. Virtual teams are particularly challenged and may not work well where the level of internal competition is high. Face-to-face collocation is sometimes the only way to resolve differences and bridge gaps.

For virtual teams, interdependence is the key measure of tasks that are the organizing focus for all teams. When the team's work is maximally interdependent, it is also most cooperative, which we label positive (+). When the work is most codependent, which we label negative (−), it is most competitive. A virtual team structures its motivations by how it chooses and designs its goals and the work that follows (Figure 6.2).

- At the top of the scale, great teams working together can achieve great things, "big win/wins." Here we see the greatest payoffs

Figure 6.2   Goal Interdependence Gauge

			Goals	Rewards
Group Interdependence	+		Cooperative	Big Win/Win
				Win/Win
Individual Independence	0		Independent	Individual Win or Lose
				Win/Lose
Group Codependence	–		Competitive	Big Lose/Lose

from the synergistic effects of cooperation. When people from diverse specialties share information for complex work, they experience interdependence that is mutually beneficial. Often, work runs ahead of rules and procedures into areas where few criteria are available, and the group must develop its own rules through participatory agreement.

- As we move toward the middle of the scale, less intense interdependence and cooperation move to more of a win/win reward system. There a mix of individual efforts combined with cooperation to solve joint problems determine success.

- Individual tasks sit at the midpoint of the scale. Here work is unrelated in terms of success or failure, reward or punishment. At this middle extreme, tasks are *relatively* simple, requiring bounded knowledge and little information exchanged with others.

- Moving down the scale, self-reliant independence turns more competitive. Scarcity structures the reward system. I win, you lose, or vice versa. Good things, like salaries and promotions, are in short supply and the best will rise to the top by winning over others. So goes the conventional logic of motivation in hierarchy-bureaucracy.

- At the bottom of the scale, competitive codependence intensifies in a battle over common resources in a zero-sum game. The larger the competition, the greater the potential loss to an extreme of mutual destruction where win/lose utterly annihilates in a lose/lose fireball. Tasks here are complex and group oriented, but people use information as a weapon rather than share it as a common resource. Rules need to be strong and enforced with disagreements resolved by the hierarchy if necessary.

"Me" or "We" Tasks

Tasks are where the virtual team's goals become real. The actual work of the group defines how interdependent it is. *But not all tasks are group tasks* (Figure 6.3).

Groups offer no advantage for tasks that have a "right" or "wrong" answer. The smartest or most knowledgeable person in the group almost always provides the correct answer. Individuals are the best performers

Figure 6.3   Group/Individual Task Gauge

		Information	Criteria
Group Interdependence	+	Complex Info Sharing Needed Diverse Specialties	No Criteria Available Make Rules Participatory Agreement
Individual Independence	0	Simple No Info Sharing Single Specialty	External criteria "Right/Wrong" Answers No Agreement Needed
Group Codependence	–	Complex Info Conflict Diverse Interests	"Might Makes Right" Rules Enforced Agreement Imposed

on tasks for which there are external criteria and standards of judgment that require no agreement to validate the outcome. Individuals are also most efficient in completing simple tasks. Individuals are particularly appropriate for tasks that require only a single expertise and no information sharing.

Groups perform best when there are no obvious right or wrong answers, no convenient external authorities to validate decisions using an impartial standard of truth. Complex tasks are the province of teams, particularly where diverse information needs to be integrated. Groups are indispensable where tasks such as innovation depend on information sharing. Even under conditions of competition, only a group can perform tasks that require interpersonal agreement such as negotiations or conflict resolution.

To some degree, the individual or group issue is just a question of scale. Human civilization was founded on the ability to do things together that individuals acting alone could not. Companies are formed to harness the energies of groups of people to do bigger, more complex pieces of work that are beyond the ken of individuals. As big jobs break into little jobs, tasks generally get pushed from the group domain to the individual one.

This is the crux of managing the core task strategy of virtual teams. This is the level at which work becomes defined. Traditional management attempts to break work down as logically as possible to individualized pieces. Most tasks are designed to be performed by one person working alone who is measured against impartial pre-established criteria (quotas, targets, benchmarks, "making your numbers"). In today's complex, fast-changing world of work, this level of routine micromanagement is increasingly untenable. For virtual teams, it is impossible.

The easiest way to ensure that a group's task will be interdependent is to back up a level: Give people a mission and leave the "task" of becoming a team up to them. Charter the team to determine its own goals, tasks, and results required to deliver the bottom-line outcome.

For most people, work life is a mixture of independent tasks and interdependent interactions—working alone and working together in patterns that vary by the hour, day, and season according to the nature of the job. Teams vibrate with a pulse of aggregation and dispersion. This pulse permeates the work design either intentionally planned or because of the accidental reactive unfolding of events.

The Strategy of Cooperation

Although we have stressed the benefits of cooperation over competition, these two fundamental tendencies in life are in a dance with each other. "Co-opetition" is the newly coined term for this uneasy dynamic of simultaneous cooperation and competition.[14]

While competition and cooperation are complements, they cannot be evenly matched. If they are, progress stagnates and change dies. One tendency or the other must dominate to carry the process forward. In virtual teams, cooperation provides the greater driving force.

> Cooperation is the fitter survival strategy for virtual teams. When necessary the smart cooperator is also an excellent competitor.

Cooperation sounds nice in theory, but should we heed the conventional wisdom: "Nice guys finish last?" Apparently not.

The tooth-and-claw Darwinian competition that many assume to be the natural condition of life is giving way. There is accumulating evidence that cooperation is evident at all levels of biology's kingdoms—from cells to big-brained mammals. It may be particularly evident in humanity's remarkable spurt of evolution over the past few million years. Cooperators seem to be the survivors.

Game theory, a mathematical discipline that explores the relationship between cooperation and competition, originally proved the futility of cooperation. The second generation of game theory demonstrates the powerful logic of cooperation and why it is an even stronger survival strategy than competition.[15]

In the original logic of games, an aggressive competitor invariably won over a willing cooperator because they only played single games, one at a time. However, if the game expands with more rounds of play involving more people, behavioral consequences change dramatically. When the news of people's behavior in past games becomes available for future games, it carries a self-correcting social consequence. If you do in another person and no one else hears about it, you can probably get away with it. Yet when such behavior becomes public and the basis for future interactions, others will not want to play with you.

The reasoning is common sense. If people know me to cooperate, they will associate with me, and together we can do more than we can separately. Cooperators win.

Perhaps the most famous event in game theory history clinched this view. Robert Axelrod, a leading practitioner of games, staged a competition to find the best strategy that logically combines competition and cooperation. People proposed various strategies that were translated into lines of code. These were in turn put into the equivalent of an open cyberspace market so that games could undergo many iterations. Anatol Rapoport, the mathematician who was one of the original four founders of the Society for General Systems Research, submitted the original winning strategy. It remains the undisputed champion. With both a catchy name and the shortest code, "Tit-for-Tat" is simple: Cooperate on your first move, then match the other player's response with the same strategy. You might call it "tough cooperation." In short:

Reach out, then respond in kind.

Open with friendship then respond to opportunities with cooperation and challenges with competition. This strategy works even where there are only a few cooperators in a sea of competitors. Tit-for-tat cooperation will slowly accrue benefits while competitors can at best achieve a standstill as they beat up on each other.

The advantage of cooperation will only grow in the years ahead. At the same time, the payoffs from purely competitive strategies likely will diminish. In the age of information, the foundations that support competition are shifting dramatically:

- From material scarcity to information plenty;
- From limited information to information access; and
- From anonymous players to trusted partners.

To cooperate and gain collective advantage, people must come together somehow somewhere. A virtual team must create a place to carry out its process.

CHAPTER 7

VIRTUAL PLACES

Home Is Where the Site Is

In the last decade of the millennium, the transition to the Information Age is in its tumultuous "storming/norming" phase. Communism crumbled and market forces reign. A new economic world order prevails as nations scramble to stabilize political patterns. The proliferation of computers and networks approaches critical mass as electronic technology shifts from analog devices that reproduce information to digital designs that enable infinite mutations. Electronic bits now drive the great transforming processes of an increasingly ascendant information-based civilization.

Virtual teams are little glimpses of the future, experimental triads of people, organization, and technology. Living a vital part of the future today, virtual team members are imprinting a bit of themselves on the shape of things to come. Their failures and victories inform successive teams and networks. Of considerable consequence is how virtual teams use digital technology. One stop on this journey is a company that lives by the technology it makes and that launched 70 virtual teams all at once.

SunTeams: Increasing Customer Loyalty

In 1993, Sun Microsystems, the Silicon Valley maker of network computing solutions whose motto is "The network is the computer," began to focus on quality in a unique way. "We asked ourselves how we could embed quality into our corporate DNA," says Jim Lynch, Sun's director of Corporate Quality. To address the question, Scott McNealy, the company's CEO and one of its four 1982 founders,[1] convened a series of annual meetings for his senior staff. Chief executives of other leading companies were featured speakers. In 1993, Federal Express' CEO Fred Smith addressed the group, followed by Motorola's CEO Gary Tooker in 1994 and Xerox's Paul Alaire in 1995.

Two major themes surfaced from these yearly meetings: Each CEO stressed the importance of teamwork and recommended getting employees directly involved in customer satisfaction.

Tooker's address particularly resonated with McNealy. Pointing to the significant impact of teamwork on Motorola, Tooker provided McNealy with the outlines of a model that Sun would follow. If Sun could apply its extraordinary technology strength to resolving its quality issues, it would be ready for the 21st century.

The classic "lean-and-mean" company, Sun had always celebrated the independence and initiative of its individual engineers. Lynch describes them as "bright engineers walking to their own drum beat and reinventing the ground rules of computing." Curt Crosby, who coordinates the team effort for Sun Microsystems Computer Company (SMCC), the company's largest operating division, which designs and manufactures its products, describes the culture as "the basic hero mentality." Thus the move to encourage teams required ingenuity and a particular spin that would appeal to Sun's free-wheeling culture.

There had been teams before at Sun. "We have always had a lot of teams self-forming in the natural course of doing their work," says Lin Brown, SMCC's quality director. What was about to happen at Sun, however, was something new—the intentional use of cross-boundary teams to tackle the company's most challenging issues.

"When you're tampering constructively with a company's DNA, you have to be very careful," says Lynch, who was the architect of what would become SunTeams, a new companywide initiative launched in 1995. Sun's strength is technology innovation. This meant that the company had to execute its team-based drive toward customer and process improvement with great delicacy. Lynch points to Java, Sun's paradigm-shattering innovation that delivers chunks of software over the Internet as needed. "It was not a technology idea that came about because we were improving processes." Sun's consequent challenge was "to keep the best of what we've got and improve what needs improvement. It's extraordinarily complex."

Jump-Starting Virtual Sun Teams

Once McNealy and his staff decided to launch SunTeams, they moved quickly. In September 1994, just a few months after Tooker's visit, McNealy and his staff met with leaders of Motorola's team effort. With more than 5000 teams operating throughout its company, Motorola is widely respected as a model for teams in large firms. Because it is in a related industry, its experience seemed a particularly apt model for Sun to follow. Five months later in February 1995, Lynch got the go-ahead from McNealy and his staff to implement "A SunTeams Architecture."

"The basic idea was a 'lightweight, one-size fits all' approach that was nonbureaucratic," Lynch recalls. To keep things simple, they developed a seven-word definition that would be immediately understandable by Sun's 17,000 employees:

Process improvement through teamwork for customer satisfaction.

This purpose statement provides a high-level goal (customer satisfaction), the means (teamwork), and the result (process improvement). A consistent definition is one point of Lynch's 10-point architecture for SunTeams that covered the basics of high quality and good teamwork. The other points are:

- A customer, either external or internal;
- A common methodology;
- Continuous and sustained process improvement;
- Education in SunTeams-specific courses at Sun University;
- Management support (which was required);
- A one-page start-a-team process;
- Annual team recognition;
- Team rewards throughout the year; and
- A consistent set of criteria for measuring team effectiveness.

To kick things off, McNealy took to The SWAN (Sun Wide Area Network) in April 1995. The company's vast computer network includes "WSUN Radio," not literally a radio station, but rather an internal Web site that transmits text, graphics, audio, and video. McNealy's kick-off broadcast was the first of three. He took to the airwaves again during the summer of 1995, then reinforced the message for a third time in October 1995.

McNealy encouraged people to become involved in SunTeams—and proffered an appealing incentive: The first annual celebration would take place in San Francisco the following March. The 16 finalist teams from across the company, selected from the major divisions, would attend "SunTeams Celebration 1996." Members of the teams that advanced to the companywide competition would enjoy an all-expenses paid weekend—with their significant others—at San Francisco's posh Ritz-Carlton.

While the SunTeam architects expected positive reaction to the idea, they were amazed at the response. The first year saw 70 teams—about twice the number they had anticipated—spring up across the corporation's six operating companies. Called "OpCos" in SunSpeak, they include SMCC, SunExpress, the aftermarketing company, SunService, the company's service arm, SunSoft, Sun Microelectronics, and JavaSoft, the newest OpCo.

All of the teams were virtual in some respect. Typically team members were in different locations and time zones—at minimum American East and West Coasts. They were specialists in different areas, such as operating systems and networking experts. Not infrequently, they came from outside the company altogether: Suppliers and customers were members of numerous teams.

Shrinking the Dissatisfiers

Through its research on quality, the company had identified 32 "customer dissatisfiers," such as late delivery of products and slow response to customer problems. "SunTeams are working on things that are important to us, not redesigning the lunchroom," Lynch explains. By aligning the Sun-Team effort to initiatives that addressed its customer satisfaction drive, the company virtually guaranteed that the teams would have an immediate impact on company performance.

SunExpress' Customer Order Cycle Team (see Chapter 1) developed an entirely new EDI-based (Electronic Data Interchange) system. It allows major customers to place their orders online and receive them within three days with minimal human intervention. The 15-member team was cross-functional (operations, marketing, sales, information resources, and finance), cross-geographic (Massachusetts, Illinois, Texas, California, Japan, and The Netherlands), and cross-company (including both a customer, Motorola, and a supplier, Caterpillar Logistics Systems).

Amazingly, the team successfully completed its work within seven months without ever meeting face-to-face. "We never had the entire team in the room at the same time," says Bill Crowley, operations manager-North America for SunExpress, one of the team's two co-leaders. Instead, the group held two-hour weekly conference calls with as many people as possible gathered together around speaker phones in their locations.

SunService's Live Call Transfer Team, based in England, significantly reduced its customer response time by entirely overhauling its call answering process. The redesign involved everything from creating new office space to installing new telephone technology to crafting new job descriptions. While most of the team members were collocated, they came from separate functions and worked locally on a 24-hour clock responding to customers in all time zones. Because they could never take everyone off the phones for a meeting (after all this was the group responding to customer problems), most of their communication was by e-mail.

As with many of the SunTeam efforts, the Live Call Transfer Team's success has led to a new initiative, the "7 × 24" project. "We can't have specialists in every country," notes David Gibson, who managed the Live Call Transfer Team. "We're designing a system that will allow us to offer 24-hour global support independent of where the engineers are located."

In less than a year, SunService's Two-Day Customer Quality Index (CQI) Team radically improved the rate at which they resolved customer problems. Before the team's work, it settled 54 percent of the problems within two days. After the team did its work, the number jumped to 70 percent. At the same time, they cut the backlog of open customer problems (what the Sun folks call "train wrecks") by 49 percent in just nine months. By July 1996, the backlog had decreased to barely 25 percent of the problem at its zenith.

"It was a grassroots team of about 16 people who interviewed all 340 people in our service center to find out the sources of the problems," says Tom Young, SunService's Customer Service manager who led the effort. "Every technology that we touched on had someone working on the project." Team members came from five different engineering groups located on the U.S. east and west coasts.

The Reliability Management System Team comprised 25 members from SunService and SMCC, 12 functional units, and three continents (Asia, Europe, and North America). It tackled a problem of such proportions that it has re-upped for the next year of SunTeams. "We started from scratch on something that is huge for the company," explains SunService's Worldwide Quality Program manager Celestine Lee who leads the team.

The problem that the RMS team is working on is how to provide integrated processes, metrics, and tools for detecting and resolving product incompatibilities once they are in the field. In this case, the field is global because Sun sells its products worldwide. "Our systems go down to the lowest component—major subassemblies such as [logic] boards, power supplies, and monitors—and we need to know how they operate across a number of different platforms. A [disk] drive may work fine in one platform and not as well in another. We need to be readily aware of the problem, get to the bottom of it, and resolve it as quickly as possible," she explains. Among the team's first "products" was a metric for detecting subassembly incompatibilities that people could understand readily worldwide.

"Now we have the additional challenge of existing beyond the 'normal' life span of a SunTeam," Lee observes. "We have to continue to evolve the team, ensure that it sustains trust, and find ways to maintain momentum."

The impressive results of these teams' work are typical of most of the efforts. Though a few of the teams foundered, none was a categorical disaster. With such positive results, the company is expanding the effort. "From a SunTeams perspective, we're on a roll. We're all fired up but check in with us in three or four years," Lynch cautions. "After the first celebration in San Francisco, no one doubted that it was the beginning of a new era, but everyone also understands how complex this is."

"We've Done Away with Paper"

Three aspects of Sun's virtual team program merit study by other companies because they are beacons of virtual team success: sponsorship, preparation, and infrastructure.

First, Sun insisted that every team have an executive sponsor.

Sun did this right from the start using their peers to introduce the idea to the senior executives. "We took a lot of our ideas from Motorola and Xerox," observes Lin Brown, SMCC's quality director, "which laid the groundwork for top level buy-in. The executives were committed from the beginning."

To ensure ongoing executive involvement, each team had to recruit an executive sponsor. "The executive sponsor is important even for simple things such as approving travel budgets," says SMCC's Crosby. "The teams could just decide to get themselves together and do it, overcoming the first-line manager resistance to spending money on travel."

SunService's director of the East Coast Solutions Center Scott Woods, who served as executive sponsor of the Two-Day CQI Improvement Team, agrees. "Executive support is needed because the team leaders don't control all the resources and budgets. The sponsor has to make sure the people who do control the resources understand."

McNealy's staff stayed involved; they were the judges for the final San Francisco competition, which proved to be highly significant for the attendees. "A lot of them had never seen these executives before," Brown recalls. "They were people that you hear about but never see. It was a really big deal to present to the Executive Management Group and get to socialize with them."

Second, Sun prepared carefully for the SunTeam launch while leaving room for a great deal of flexibility and creativity.

Unfortunately, many companies decide to move to teams without a great deal of forethought. Often an edict comes down to "form teams," with no supporting guidelines. Equally frequently, a company launches its team initiative with so much bureaucratic baggage that the effort is stillborn before it begins.

"One of our themes in SMCC is to have a very skeletal foundation so that the teams can go off and run with it in each of their organizations," says Crosby. "While we have to put some fundamental processes in place, organizations can creatively add to the process." This means that each team is free to develop its own agenda and schedule while holding administrative overhead to a minimum. Virtually every team that applies to be a SunTeam is accepted. When teams experience unanticipated conflicts, they quickly resolve them themselves with guidance from the team sponsors.

Third, Sun had the technology infrastructure to support a large number of virtual teams.

Sun has been a boundary-crossing e-mail culture since it began in the early 1980s. "The Internet has always been the backbone to Sun's approach to computing," Brown points out. "We use e-mail for everything that people in a lot of other companies use the phone for."

Possibly connecting the world's largest intranet in terms of Web servers, SunWEB did not even exist in 1994, but by 1996 it had 3000 servers connected to it.[2] "We started using the [World Wide] Web to support cross-boundary work the moment Mosaic [the first graphical browser] was discovered," Brown says. "It was a real natural for us. Now we handle an incredible number of things over the Web: internal employee handbooks, manager handbooks, benefits information, quality data, and all kinds of tools. It's become our method of choice for internal communication. It's so easy and effortless. You can take any piece of information and put it on the Web in about 10 seconds. We've done away with paper and moved to the Web."

At Sun, the Web is the place.

Moving from Place to Place

"If you want to change an organization, the best lever is to change how it communicates," says W.R. "Bert" Sutherland, director of SunLabs, Sun's research and development group. "The big change of our time is what engineers call the 'time constant.' You can go around the globe in a matter of a few seconds in e-mail; the postal service takes days or weeks; in the windjammer days, it took months. A phone call is instantaneous if I can get through. E-mail is fast but not instantaneous and you don't need the recipient's attention. Different communication styles lead to different organizations."

While organizations can enormously increase their effectiveness with the smart use of technology, heed what we have heard repeatedly from our on-the-ground virtual team experts: "It's 90 percent people and 10 percent technology." Social factors above all derail the development of many virtual teams. Understanding the new "social geography" of media, as Sun is doing, provides a powerful advantage in constructing productive virtual work places.

Increased access to information is a primary driver of change from hierarchy-bureaucracy to networks. Virtual teams depend upon the open exchange of information, both internally and externally. Still there is a danger here.

Absolute openness will absolutely kill virtual teams.

As more information becomes more public, privacy becomes more precious. If all of its information and communications are public to everyone all the time, a virtual team will:

- Have more difficulty coalescing its identity;
- By-pass socialization rituals; and
- Remove essential supports for authority.

Issues of what is public, what is private, what is open, and what needs to be secure are central to virtual teams. In particular, these issues impact the design and development of cyber places, the true homes of fully realized virtual teams.

The Play Is the Thing

No Sense of Place[3] is the title of Joshua Meyrowitz's ground-breaking book exploring "the impact of electronic media on social behavior." The essential message of the book is that electronic media are dissolving the historic connection between physical place and social place.

Meyrowitz brings together Erving Goffman's concepts of how social settings influence roles with the mind-popping work of Marshall McLuhan who described media as extensions of the senses (see Chapter 4). Communications technology sets the stage for a whole new roster of roles as place expands into the ether.

Goffman said each role has two sides. Using the metaphor of a play, he described the role as presenting its public face to the audience and its private face "backstage" where the actors and director develop, rehearse, and discuss performances. Historically, belonging to a group has meant being able to go backstage. New people socialize into the group through their gradual introduction to the backstage. There they gain "inside" information. Promotion in a hierarchy means moving to ever newer, more exclusive stages.

Since time and place have historically been coincident, Goffman simply assumed the obvious, that groups communicated primarily face-to-face. Until now the more subtle relationship between physical space and social effect has been obscured.

"It is not the physical setting itself that determines the nature of the interaction, but the patterns of information flow," Meyrowitz writes. If the social setting is an information system, then new media dramatically change the roles that people play. He places roles in three categories essential to virtual teams: identity, socialization, and rank.

Identity

For the group to have its own unique sense of *identity*, its physical location matters less than the "shared but secret information."[4] Members have access to this privileged information where and when the group gathers, providing them with a core sense of belonging. Shared but secret information separates members ("us") from others ("them") who do not have the same access. Backstage the team discusses options, resolves conflicts, and makes decisions.

Suddenly, in the electronic era, people no longer must gather in physical places to "belong." Virtual teams tend to have very porous boundaries and may have little or no backstage. As private group places become public ones, group identity, an elusive quality hard enough to establish in the virtual world, blurs.

Socialization

New people become members of a group through "controlled access to group information," the formal and informal processes of *socialization*. Orientation and training are formal processes of socialization, while hints, tips, and suggestions convey crucial knowledge informally. People grow into groups over time. When access to a physical place governs availability of information, the whole group can watch as new members transition into full participants through their rites of passage.

Since it is physically impossible to be in two places at once in the face-to-face world, access to new places also used to mean that you had to leave old places behind. The electronic era suspends the Newtonian laws of motion. Here people do not have to desert old places in order to access new ones. You can simultaneously attend numerous online places, acculturating yourself to new groups while weaning yourself from old ones. You even can multiply synchronous interactions: One European member of a major U.S. corporation's executive committee attended one of the group's meetings by video conference. At the same time, he took phone calls and talked to frequent office visitors. Where exactly was he during the meeting—or was he attending multiple meetings simultaneously?

As physical places give way to virtual ones, new members can instantly gain access to all of the group's information. Not surprisingly, traditional patterns of socialization are collapsing as transition stages become more difficult to discern.

Rank

According to tradition, *authority* is highly dependent on access to exclusive places that house special knowledge. Elite clubs are obvious locales that demonstrate the power that comes with place. University libraries are another; if you belong to that particular academic "club," you have access to its special knowledge which can literally make you an *authority* on a subject.

Indeed, the higher the group is in the hierarchy, the more these socially remote places convey a sense of "mystery and mystification."[5]

Inaccessibility is a measure of status (or lack thereof). Members jealously guard backstage areas and carefully script performances.

Since the Nomadic era, new media have increased the ability of leaders to segregate and isolate information systems. The consequence is the extension of control. Here again, the electronic era is chipping away at these bastions of privilege. While it still may cost many thousands of dollars to join the country club, you need only pay your monthly Internet provider fee to enter into conversation with countless numbers of experts everywhere in the world.

Likewise, anyone with a modem and a World Wide Web browser now can visit thousands of university library home pages without ever registering for a single university course. Yet if that same person showed up at one of these libraries without an official identification card, access would likely be denied.

Another irony of the electronic era is that an anti-status symbol of the past is now an important tool to sustain authority in the future. Typing, once considered the province of the hired help, is a key skill in the electronic world. The effect of broader access to once-exclusive information has been felt nowhere more profoundly than in the upper ranks of hierarchy.

The "Construction" of Virtual Places

The need for some degree of privacy is one of those archaic features of groups that remains essential for virtual teams.

> *Privacy complements openness as individuality complements group cooperation.*

In general, virtual teams face more hurdles in establishing their identities than do collocated ones. Shared, exclusive information is one way that a team develops a strong identity. For many groups, privacy is essential. Such is the case with Buckman Laboratories' (see Chapter 2) online Research and Development discussion area where patentable products are under development. "Inviting someone into that forum is asking someone to look at your research notebook," says Victor Baillargeon, Buckman's former vice president of Knowledge Transfer.

Corporate borders secure the absolute need for some information exclusivity in the competitive private enterprise system. Membership and privacy are invariably established at the enterprise level. There an account on the corporate information system accompanies the badge with a picture for access to the physical facilities. At Buckman Labs, "membership" as an employee in practice means an account on CompuServe and passwords to Buckman's online discussion areas. Some of the discussions are open to the entire company and others are restricted.

For decision-making and negotiating tasks, team privacy is essential. Openness to disagreements and an ability to tolerate yet manage conflict are at a premium in healthy boundary-crossing groups. Yet these qualities are even harder to foster in a fish bowl. The 10-minute video of Sun-Teams preparing for their final presentations for the competition in San Francisco contains several amusing scenes poking fun at their need for privacy. Teams rehearse in private and present in public.

It is easy to design digital places that combine public and private areas, most simply through passwords and access lists. We have already noted that virtual team boundaries tend to be multilayered. Often they comprise a small core group, an extended team of less-directly responsible members, and an even larger network of external partners and tangential people. Companies regularly configure multi-level virtual spaces. Internet sites allow public access to published information, such as press releases and annual reports. Internal intranet areas require authorization with access to plans and interim results. Completely private places are where teams discuss their most sensitive issues, such as budgets and personnel matters.

By creating information places with graduated levels of access, virtual teams more easily and naturally stage the socialization of their members. At Buckman, for example, new employees begin by perusing the generally available information as a way to get to know the group's public persona. Soon, they receive passwords that offer access to the "regular" inside information of the company's work. Later they are invited to join certain discussions with information that is proprietary to the group.

Virtual Ladders and Competency Networks

The social effect of increased access to information is most dramatic in the shrinkage of hierarchy—which is flattening but not going away. For the most part, middle and supervisory management ranks are dwindling. Executive management is, if anything, becoming more exclusive and remote, a trend symbolized by the steep increase in CEO salaries. For all the personal aversion of many senior managers to computers (a dying generational artifact), the best and most powerful tools of digital technology have always been put at the service of executive information systems. This is not likely to change in virtual organizations.

Executives face the greatest challenge in making virtuality work for themselves. They above all must balance two apparently conflicting needs. On one hand, they must follow a general admonition to share information cooperatively and broadly throughout the organization. On the other hand, they have the strong requirement to protect the privacy of their own deliberations and "below the waterline" information (the disclosure of which might "sink the corporate ship"). The behavior of protecting exclusive information from subordinates is all too easily carried into executive team relationships. One unfortunate consequence is a corresponding diminution of cooperative pursuit of overall corporate goals.

Paradoxically, while hierarchical *boss*-ship contracts, virtual teams and networks demand more *leader*-ship not less. Many leadership roles are changing. Virtual team leaders often act more as coaches than bosses. They are more likely to lead through influence than coercion, and are much more diverse in their sources of power.

Like vertical leaders, horizontally linked leaders need their private places. They too exchange peer-related information, debate standards, criticize rules, challenge orthodoxy, and otherwise prepare to meet their public leadership tasks. Membership in competency groups is usually "by invitation only" based on expertise and/or position.

> *Competency networks that link people with common expertise (such as technical) or similar roles (such as project managers) address the need for horizontal leadership in virtual teams.*

Where Place Is Going

Metaphors from the physical world regularly tag the online one. People sitting at computers work on their own desktops while accessing group information on servers at sites. Desktops may be a metaphor that in time will seem as quaint as horseless carriages. Regardless, some sense of place—like a site—will persist in the human online experience.

Site is a cross-over term. It simultaneously stands for a building (or group of them), a computer or a cluster of machines, and an ephemeral place of bits in cyberspace, as in a World Wide Web site. Physical and online sites alike range in size from small to gigantic. At the small end of the scale are physical and online "rooms." At the other end of the scale are corporate campuses like Microsoft's in Redmond, Washington and vast cyber facilities like America Online.

As teams and organizations expand their presence online, they will continue to create online places that are analogous to the information resources in their physical places. Each organization that goes online invariably creates its own digital place, stocking it with information and products previously available only in physical places.

The United Nations Development Programme (UNDP) is perhaps the most electronically sophisticated group at the global organization. It uses electronic networking both to carry out its mission—to build more sustainable livelihoods for all—and to encourage more direct individual and community participation worldwide in the UN. For the 1995 Fourth World Congress on Women in Beijing, John Lawrence, principal technical adviser at UNDP, and his colleagues "worked from behind the scenes," he says (echoing Goffman's language). Supported by the Education Development Center of Newton, Massachusetts,[6] the group "rented an electronic virtual room where anyone could come in to discuss issues that were related directly to agendas raised." During the summit itself, they scanned relevant documentation on to the Internet as it became available. Annotated summaries of sessions were available during or just moments after events took place so that anyone anywhere in the world with Internet access could view them.

People create online places from the ground up. To do so, they use virtual analogs of desktops, rooms, offices, factories, malls, and communities. These and other familiar "place" metaphors serve as the building blocks for local cyberspace. We anticipate these metaphors will rapidly evolve from cartoonlike storefronts and graphical menus to increasingly sophisticated three-dimensional virtual realities that members will "walk into and fly around." As the early generations of kids growing up with computers mature, they will incorporate the representational features of game technology into virtual team interfaces.

Product Places

Insofar as they could be developed in digital form, Information Age technology products always have occupied a privileged position in the world of virtual work. They benefit from a basic axiom of "going virtual":[7]

Digitize early and often. Start your results in digital form and keep them digital as long as possible.

The development of products in digital form offers one significant way that virtual teams can go beyond physical place metaphors. This capability has been slowly developing for the past two decades.

The Result Is Where We're At

One early case of an astonishingly successful global virtual team was Digital Equipment Corporation's Calypso Project in the mid-1980s.[8] This team created a revolutionary new minicomputer design. It was so robust that it served as the basis for a major product line, the VAX 6000 series. At the same time, the Calypso team built a production capacity that saw the first machines roll out simultaneously from three plants separated by an ocean. Everything was done in record time, and the project generated $2 billion in revenues the first year, and many billions in the years to follow.

From the beginning, Calypso put its whole product design online. Thus it closed the loop on what had been a gradual transition through the 1970s in engineering and manufacturing design from analog to digital processes. The project's most intriguing technology innovation was its product database that contained everything from chip design to the metal "skins" of the machines. The product design was the team's "place." Everyone on the team had access to the whole product database. At the same time, the communications system was designed to notify people only when changes were made in areas that they had previously specified as important to them. Thus, the product itself in its digital form became a highly specialized primary communications medium.

While a computer design eventually must go from bits to atoms as a machine is made, software is a pure product of the digital age. Software is a truly ephemeral "thing" that naturally lives in virtual space. Software teams have always been at the leading edge of virtual work. Two key factors genetically code them for success. First, they have a commonly accessible online product focal point for their interdependent tasks. Second, they tend to have the necessary computer technology for communicating easily across boundaries. In our experience, the weakness of distributed software teams usually lies in their people and organizational issues, not access to their common product or the availability of technology.

One early very successful global software project was the team that developed the Ada language. Military and other applications that require very fast real-time data processing for systems such as the Boeing 747 use Ada. Beginning in the mid-1970s, a core group of a half-dozen people engaged with a larger set of 100 key contributors in 20 countries. Together, they carried on a complex set of technical conversations over the DARPA network, the military forerunner to the Internet. Over the multi-year course of the project's development, the conversation volume grew to 10,000 comments.

As with Calypso, the Ada product was both online and shaped the team's (online) conversation about it. Jean Ichbiah, the Ada project manager and now CEO of Textware, credits the early establishment of a coherent architecture (that is, strategy) as the key to organizing the talents and time of the larger team of teams. The architecture and the creative issues it posed provided the classification system of topics that structured conversations among team members. As conversations came to resolution, results would accumulate in the language product. Ichbiah believes that "distributed product development is very positive because it requires the process to be more structured and formal, with well-defined interfaces between relatively independent components."

"The story was very interesting from a networking standpoint," says former Apple senior vice president Ike Nassi, who was originally a reviewer of the Ada over a five-year period. "Remember: this was DARPANET in the very, very early days. The reviewers worked with the language design team that was drawn from groups in many countries. We had a series of very official language design notes that were issued by the language design team with extensive commentary. It was a very formal process. We'd download the notes and then send lots of e-mail back and forth. It was almost Talmudic in nature. Visualize scholars sitting around a virtual table pouring over scrolls and arguing over Judaic interpretation. A lot of thought went into a lot of issues and in the end Ada popped out."

All Virtual Presence

Although they were pioneers in complex virtual team collaboration, Ichbiah and the Calypso team managers also attest to the importance of face-to-face meetings as a necessary part of the communications mix. In "extreme virtual teams," however, face-to-face plays little or no role.

Lynx is an example of a very large-scale, completely voluntary distributed software project community that operates with very little face-to-face contact. This is the Internet-based global group of over 500 engineers and other professionals who develop, maintain, and evolve Lynx, the Netscape of text-only Web browsers.

The Lynx network is organized into a teamnet of specialized working groups that use the simplest form of digital interaction, an e-mail list. "The mailing list serves to collect code patches and to return glory to those who contribute them," says Al Gilman, who keeps a FAQ (frequently asked questions) for the list. "It also collects trouble reports and carries discussion among the participants that are generally related to Lynx. The list participants function as a self-managed team to repair and improve the Lynx product."[9]

Using your result as the lodestone for place does not have to be big and complicated. It can be as simple as a memo or report. At the University of Texas, Kathleen Knoll and Sirkka Jarvenppa conducted studies of virtual teams who never meet yet who must produce common products. They analyzed data from 19 teams numbering from three to seven graduate students each at 13 different universities in nine countries who only used e-mail to communicate.[10]

The best predictor of success for these extreme teams seemed to be a decision "during or soon after brainstorming, to work from a common document summarized from everyone's comments. This process seemed to help the teams collaborate." Teams with a common document early in the process generally communicated more frequently. They also had more consistent and even participation, showed less conflict, and evinced more satisfaction in the project. Finally, they demonstrated a greater "sense of team," meaning that they communicated "feelings, context, sensory information, roles, and identity."

Virtual Technology Principles

As place becomes ephemeral and moves online, it also has a physical existence in technology. Network technologies and organizations are co-evolving, each influencing the other. The principles of distributed organizations complement the principles of distributed technology. Virtual teams are small group networks coming to life in the age of computer networks, and vice versa.

A severe organizational dissonance arises when a company installs new network technologies without changing its traditional hierarchical-bureaucratic management. The then-dominant form of mechanistic organization shaped early computing with its massive mainframes and "slave" machines, totally dependent "dumb" terminals. Nowhere is this more clear than in traditional MIS (Management Information System) departments. Such centralized facilities sprang up to manage mainframes. Today the decentralized network paradigm drives computing. Virtual teams and network organizations at all levels leverage this technology best.

A network of computers and a network of people share some common conceptual elements:

- People are nodes;
- Links are links; and
- Purposes are applications.

The people/links/purpose model of virtual teams fits the features of the digital workspace that network technologies create. A schematic of a technology network often uses circles and lines. Circles stand for nodes that are individual machines, sites, or networks. Lines are the technologies that connect the nodes. A picture of people networks looks similar. Typically it consists of circles that stand for people or organizations with connecting lines indicating relationships. Often missing in both types of these circle-line diagrams is the third critical element: the purposes—the applications of the network.

The virtual team principles can help you shape networking technology to support your boundary-crossing groups (Figure 7.1).

Network Nodes

Just as the word "people" in virtual teams comprises three principles, so does its correlate in technology networks, "nodes." A virtual team's technology network has:

- Independent nodes;
- Shared servers; and
- Integrated levels.

Figure 7.1 Virtual Team and Technology Principles

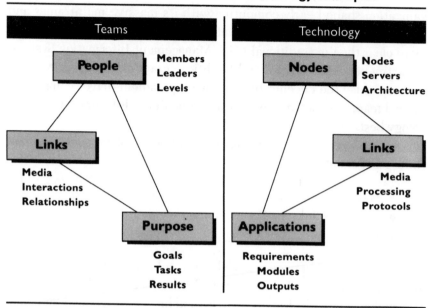

The power of the team relates strongly to the adequacy of technology tools available to the people to do their work. For virtual team members, this usually means personal computer power—a node. "Independent nodes" on the computer network enable every virtual team member to have the ability to work independently *and* interdependently. Ever since the earliest configurations of terminals and mainframes were eclipsed (which is not to say they have disappeared), computer networks have depended upon some minimal level of intelligence, or independence, in their nodes. When your LAN or Internet connection goes down, your PC, your independent node, continues to function. Ideally, all virtual team members have their own PCs.

Computer network servers reflect the shared leadership of virtual teams. Servers house information that is common to virtual teams—their databases, applications, documents, and other files that belong to the group as a whole. In the client-server computing paradigm, typically each organizational unit has its own server, for example, a departmental server.

Virtual team leaders often have responsibility for the information on the server and for who has access to it. Servers suggest *service,* an important metaphor for the emerging mode of leadership in virtual teams, what Robert Greenleaf calls "servant leadership."[11]

Just as virtual teams, teamnets, and other people networks are structured in levels, so are technology networks. From the ever-shrinking gates on chips to processors and peripherals to the awesome reach of networks of networks, complex computer and communications systems follow an inclusive sets-within-sets-within-sets design. A multilevel architecture is a *sine qua non* of good digital technology. For virtual teams, integration is a practical matter. It arises both in the technologies the team chooses to support its functioning and in the technologies it uses to generate its work product. Will our applications work together? Can we link to the corporate networks? Can we connect remotely? These are key pragmatic issues that virtual teams must resolve.

Connecting through Links

Organizational links are the great differentiators of virtual teams—supported by their corresponding technology links:

- Multiple media;
- Boundary-crossing processing; and
- Trusted protocols.

Nodes alone do not a network make. Both technology and people networks need physical links. Virtual teams access multiple media drawn from the array of communication forms developed over the ages. For interactions other than face-to-face ones, technology links are the physical connections that virtual teams live by. In the digital era, the computer-based medium includes all previous media. It is increasingly easy and commonplace to mix print, audio, and video into multimedia. Even handwriting plays a role in virtual teams: People share virtual whiteboards through their Web browsers while working synchronously with audio and sometimes video links. (Still nothing has quite the feel of a handwritten personal note, increasingly cherished in the digital era.) The best virtual teams use multiple media.

Virtual teams need to be able to move their work across complex and diverse technology boundaries. For virtual teams, processing leaps across space, time, and organization boundaries. Data flows to and from and among many different locations. More than one virtual team member has lamented, "I couldn't read your file. It came across as gibberish." If a team cannot move its work from node to node, it clearly impairs its ability to function effectively at a distance. To develop distributed work processing, you need to be concerned with the computer processing capability of your teammates as well as your own.

Trust has its correlate in technology networks in protocols. Processing across boundaries requires more than physical connections. Protocols, permissions, and open formats are based on agreements and decisions of people who manage systems. It requires people's agreements to make it possible for cross-boundary computing systems to work together. Trust in this arena is vital. In most computer systems, people do not give out permissions lightly. Establishing and agreeing to protocols are areas of cooperation that enable the development of technologies to work together. A large-scale example of this is the TCP/IP[12] Internet standard. Without thousands of people committing to use the same technical standards, the Internet simply could not function.

Purposeful Applications

The third analogy is between organizational and technology purposes. For a virtual team, goals first surface at the beginning of the group process. In a similar way, original requirements and specifications are necessary early on for technology development. Vision sparks the need for new technology together with perceptions and anticipations of user needs. Purpose grows from the seed of vision and clarifies into a system design with:

- Cooperative requirements;
- Interdependent programs; and
- Concrete outputs.

People develop requirements for certain applications. Ideally, a group's goals, which it arrives at cooperatively, drive the development of technology to support them. Remember that for a virtual team, users include the group-as-a-whole as well as the individuals in it. The cooperative work of the interdependent virtual team members determines requirements for groupware. E-mail, for example, makes no sense in the singular. We are still in the early stages of learning how to design interfaces for groups-as-users.

Tasks, the definitional heart of teams, represent the work that unfolds from the goals. Interdependent applications need to support interdependent work. Modular software programs (for example, application "suites" such as Microsoft Office) are the design equivalent of "independence with interdependence" in virtual teams. As the interdependence of work accelerates, systems of access to programs will change. They may migrate to Java applications modules delivered by the World Wide Web, which provides a virtually universal computing platform. For the interdependent virtual team, interdependent software is key.

A virtual team's decisions about the results that it will deliver inform, if not drive, their selection of technology for creating an online product place and delivering the output. For task-oriented teams, concrete results are the bottom line, so output matters. However ephemeral or fantastic the means by which the miracle of computing is achieved, screens, printers, and other output devices finally deliver meaningful and concrete results to people. Thus, virtual teams need to think through what technology supports the results of their efforts.

Using the digital representation of the product as a virtual working space is one way a team makes its shared models explicit and extends the intelligence of the group.

Thinking Technology

Creating virtual places is initially about making adequate substitutes for physical places. This is a necessary but preliminary step in the evolution of virtual teams and networks.

New technologies are innovations that diffuse through society in a well-recognized pattern. First, the new technology develops slowly against resistance, gaining a foothold by replicating and replacing functions of older technologies. Only after an innovation establishes itself as a substitute will its truly innovative features and revolutionary effects come to full expression. Then it rapidly expands through society.

> *The cognitive characteristics of groups will blossom in the fertile soil of shared digital environments.*

Members gain more than social and task information with their access to physical places. They also use them to take possession of shared "mental" or "cognitive" models. In a direct visceral sense, people acquire a mental image of the collocated team. You can easily visualize such an image as a set of unique individuals assembled in their special place. Equally important is the model of the group's work. This is traditionally evident in the space where people do the work. Materials, tools, partial products, and people identified relative to their roles all contribute to a concrete understanding of the group's purpose and how it pursues it. All of this together becomes the shared cognitive model.

As people construct new virtual places, they embed in them their shared cognitive models—consciously or unconsciously. A virtual team does not just replicate an old physical place. It also generates a new conceptual space that has never existed before.

When virtual teams explicitly share their models, their ideas go beyond the members themselves. The Calypso product database reflected the integrated result of many people's thinking, both in its overall architecture and in the countless choices people made about their communications. Thus, a significant portion of the group's shared intelligence and ongoing thinking was expressed and retained in bits online.

Cognitive Webs

In the idiom of the Industrial Era, organizations are likened to machines. In the Information Age, both organization and computer networks are feeding off the same metaphor, the human brain/mind. Where once the extension of limbs and senses occupied center stage in the human development of tools, today digital technology amplifies mental capabilities.

The abundance and variety of the links of virtual teams are their most distinctive feature—even more so than their people or purpose elements. During the initial analog phase of computer development, physical brain analogies between corporate networks and human nervous systems seemed apt. As we rocket into Web worlds interrelated through hypertext links, mind metaphors will come to dominate future descriptions of virtual organizations.

After 10 years online, Buckman Labs is still in the early stages of building its companywide online repository. The more it puts online, the more explicit the company is able to make its cognitive models. We expect that this will be the new norm for virtual teams. As they develop their shared virtual reality with more of their information online, they become increasingly explicit about their models.

The roots of these emerging models reach back into traditional hierarchy and bureaucracy as well as cast forward into the new elements of networks.

Navigating with Mental Models

Traditional Models

- **CORPORATE** identity is an executive responsibility for the public face of the whole organization. It is often the starting point for the development of external sites on the World Wide Web.
- **HIERARCHY** is most visibly represented by an organization chart. Hierarchy is a valuable navigation tool, particularly internally and for customers. Often, however, organizations treat it as a trade secret.

(Continued)

Navigating *(Continued)*

- **BUREAUCRATIC** rules and regulations, policies and procedures, guidelines, and protocols are the recently modern models of the traditional organization. Formalities and organizationwide information collections are very transferable to the online medium, usually by converting existing processes and analog media into digital forms.

Network Models

- **PEOPLE** and organizations are identified online in directories, "yellow pages," Web home pages, and other collections of individualized information. Leaders offer their own key views of the team and its work through online announcements and pronouncements. The hierarchical design of sites and their component parts—represented in some variation of an outline or a table of contents—attests to the level structure of information.

- **PURPOSE** appears as online mental models in statements of vision, mission, and goals along with the strategies and plans used to achieve them. Hyperlinked plans are a largely unexplored but potentially very powerful form of group interface.[13] Results that use the product as an analog for place offer a final destination that makes the work worthwhile. The virtual team's output provides a very valuable and practical mental model for the work of the group.

- **LINKS** generate shared images that flow from the group's communications as well as the pattern of the ongoing conversation and information exchange. A communications model accumulates through various modes of memory that store, recall, modify, and reprocess the group's stream of consciousness. Shared calendars and information associated with meetings, events, and deadlines also help people build a common model of the group's movement through time.

Some organizations already are incorporating some of these features. The first three selections (buttons) on Sun Microsystems' internal home page offer access to information through organizational, functional, and geographic models of itself.[14] Such models serve as a group interface to its common information, whether through text, outlines, diagrams, pictures, animations, or any other representational form.

> *On the Web, people can express links and relationships in context.*

In intranets, a dynamic distributed human intelligence comes together in a context that grows with the group. With hypertext links—more of a concept than a technology—the team's ability to create and use shared cognitive models crosses a fundamental threshold. The nature of the online space is no longer primarily an artifact of the hardware/software structure of the technology. It is a matter of choice, the human intellect creating a shared cognitive space.

SunLabs like all of Sun, its larger host company, uses intranets extensively. "We're witnessing the next change in communication style," says Bert Sutherland, SunLabs' director. "E-mail is a push model; I want to broadcast to someone. The Web is a pull model; the information sits there until someone who wants it can pull it."

For millennia, new media have improved the ability to *push* information. With digital media, the historic trend is suddenly reversed. People are becoming increasingly oriented to information *pull*—seeking and finding the information they need when they need it. In a "pull model" of information access, particularly where users are both readers and writers, it is vital that everyone share common views of what information goes where when.

CHAPTER 8

WORKING SMART

A Web Book for Virtual Teams

This "how-to" chapter focuses on the many practical ways virtual teams can develop and use their combined smarts, their "group intelligence."

> *On the whole, a virtual team must be smarter than a conventional collocated team—just to survive.*

Being smarter means that the team shares its ideas freely and creatively. People think together about what they are doing. Brainstorming is one obvious way that a team engages its intellect. Every diagram a team makes, every memo written, agenda proposed, and idea exchanged—all of the team's shared interactions—combine into mental or cognitive models. As people share their mental models and test them in the environment, they collectively think up better ones. The better the shared understanding, the stronger the model. Better group models equal greater group intelligence.

All teams share cognitive models of themselves as a team, of their work, and of their environments. In most situations, these models are fragmentary and unexpressed. In the conventional, well-structured collocated team with its ever-present boss and proverbial water cooler for informal interaction, their explicit articulation often is unnecessary. Virtual teams, by contrast, lack the traditional cues, and thus need clear ways to view themselves and their work.

> *To work smarter, virtual teams need to build explicit models with common categories and the right relationships.*

Abstraction is sometimes difficult for people who prefer to communicate about concrete things. Most of us like knock-on-wood, hard reality, the "I can see it, feel it, taste it" satisfaction of material concreteness. Unfortunately, these aspects of traditional work are in short supply for virtual teams. The faster, more global, more complex Information Age world demands greater abstraction. The trick is to learn how to use abstraction to advantage.

Before continuing with this chapter, please stop to think of a team that you know well. Perhaps you will choose the most successful team you have ever been on, a team you are on now, or a team you would like to design. Use your experience as you read. Enter information or check off items (for example, "We wrote down our purpose," or "Cathy, B.J., and Ron are team members"). Leave question marks where you are not certain how your example applies, or note blanks that represent missing pieces of the process that you are remembering, experiencing, or imagining.

Awash in the flux and chaos of change, the method by which a new team takes form is not linear. It "iterates" through a series of "rapid prototyping" cycles. The team does a mental mock-up of itself.

Below, we present three phases of planning and development for a virtual team. Phase 1 is a high-level overview; Phase 2 goes through a more complete planning process; and Phase 3 offers complex virtual teams a way to develop more systematic detail.

These three cycles together provide a summary of the book's most practical ideas. Thumbnail-size versions of the book's graphical figures point to concepts elaborated earlier. Follow the page numbers to find the original figures and more on the ideas (this is a print version of a

World Wide Web link). The boxes and tables are meant to symbolize fields in a database. Imagine filling in the data. Then you have the information you need to create a plan, handbook, and Web site.

Each of the phases, outlined below, has features of the basic elements of the virtual team model: purpose, people, and links.

Phase 1

Team Concepts
Virtual Team Name
Statement of Purpose
Overall Results
Delivery Dates
Location
Key Goals
Key People List
Team Size and Bands of Involvement
Contact Information
Team Types

Phase 2

Purpose Flow
Process Elements by Goal
Task Deadlines
Responsibility Matrix
Task Leadership
Process Leadership
TeamFlow Model
Distance Gauge
Media Palette
Media Plan
Members/Media Matrix
Virtual Team Handbook

Phase 3

Virtual Teams Principles Taxonomy
"Stressed S" Team Process

McGrath Task Circumplex
Cooperation/Competition Gauge
Individual/Group Gauge
Task Factors
Media Characteristics Chart
Task Timing
Hierarchy Ruler
Network Organization Chart
Virtual Team Web Book

As you consider each phase, please keep in mind that virtual teams benefit enormously from face-to-face meetings. These are particularly important in the early phases of development. If meeting face-to-face is too costly or otherwise constrained, consider using a mix of the many interactive technologies available to virtual teams, as described in the examples in this book.

Virtual teams that follow a clear process, supported by technology that captures their work as it unfolds, will naturally and unconsciously develop a tangible group intelligence. Once we know groups can be smart, we are on our way to learning how to enhance our collective intelligence.

Phase 1: Setting Up the Basics

How does a virtual team begin? People with an idea start talking and soon a new virtual team is on its way to formation. Regardless of how it

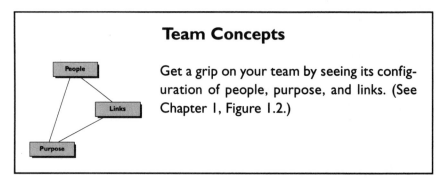

Team Concepts

Get a grip on your team by seeing its configuration of people, purpose, and links. (See Chapter 1, Figure 1.2.)

begins, a team grows in three basic dimensions—people, purpose, and links. A purpose statement and a team directory summarize key early outcomes of Phase 1.

Create an Identity

A team's name is its smallest mental model. Names may be dull but descriptive, creative expressions of mission, or wild things that capture people's imagination.

Virtual Team Name

Name yourself. A name labels the category of the team itself. Consider using a formal name that clearly communicates what the team is about, for example, SunService's Live Call Transfer Team. Use a short tag for internal use—in this case LCT—or let a nickname bubble up through the ongoing group conversation.

Names often summarize an overall purpose that a team expresses through its mission. The act of writing a vision or mission statement and then hanging it on the wall has become the well-deserved object of ridicule in many organizations. If the exercise stops there, chuckle on. When the exercise of writing a purpose statement becomes the basis for the group's work, it is a powerful source of energy. We cannot overstate the importance of a virtual team going through a process to make its

---

Statement of Purpose

Construct a prose statement of intent that answers the question, "Why are we doing this?" Make explicit the team's mission—its top-most goal and motivation to action. The purpose statement is the formal informational icon of group legitimacy.

---

purpose tangible. In the end, this means writing down the purpose and creating a charter, however informal.

Rule number 1 of every team is to get the purpose right and make it clear to everyone, a task that is at once more important and more difficult for virtual teams. Even when it receives its purpose from above, a team must interpret and express it in its own terms. In a distributed team, which functions with far less oversight than is customary in a collocated team, people must understand and commit to the purpose.

Overall Results

The primary team product is the answer to the question, "What are we going to do?" The bottom-line result is the group's *raison d'etre*. Mentally place yourself at the end of the project and then look back. Use words and draw pictures that describe the final product of the team's work.

A team often expresses the concrete image of the ultimate result of its work in numbers—more market share, lower cost, faster cycle times. Decisions, events, reports, presentations, prototypes, or anything else that represents the concrete consequences of joint effort are also examples of end-point results.

Delivery Dates

Document the team chronology from the beginning. Define any known deadlines and establish a best-guess time limit for results. Create a list of significant external dates, such as budget cycles and major conferences. These dates anchor the team calendar and begin to rough out the phases and pace of activity.

Teams live in time. They sputter into life as people talk, meet, argue, agree, and formalize. Early team history accumulates as people make contact and develop relationships. Key outside dates that impact the team help shape the anticipated timeframe. To establish the boundaries of their overall process, the team needs to set delivery dates for results at the beginning, however imprecise the estimate. They can be set as a one-time deadline, a set of milestones, or as periods of performance evaluation (for example, quarterly or annually for ongoing teams).

Location

Does the virtual team have a natural home given the circumstances? Identify physical locations and phone numbers where applicable. Set up electronic addresses and places where possible.

Location, like all these early features of virtual teams, may not be very clear or specific on the first iteration, but it will over time become a given of the team's work. Soon virtual teams will come to associate themselves with Web locations, which are infinitely configurable suites of group and individual workspaces.

Name the Goals

To get from abstract vision to concrete realization, you need to organize the work and decide who is going to do what. A deceptively simple approach to work process design is to establish a set of goals which, when achieved, together accomplish the overall purpose. Well-conceived goals become the major components of the team's work and the seeds around which subgroups form to actually do the work. These major internal team components are like the internal functions of a corporation—units of work that also identify clusters of people.

Brainstorm this list early and often, keeping the categories fluid during the first phase as purpose, people, and links are initially detailed. Look ahead to nailing them down in the next phase as the basis for the team's internal organization.

Key Goals

Make a first pass at naming the key goals of the team. Keep the major categories to a handful or two at the most. Assess whether this set of goals covers the statement of purpose and the overall result. Consider how people can work together in goal-oriented subgroups.

Identify the Players

People or purpose, which comes first? The natural impulse is to immediately come up with a list of people. The more practical approach is to first draft an initial purpose and then identify whom you need to involve.

Key People List

Name	Organization	Location
...	...	...

Figure out which organizations you need to involve in your project, then identify specific people for the team. Note the organization of everyone involved and the location of their primary workplace. This is the starting point for the team directory.

A team emerges from the activities of specific people who can be represented by a list of names. Early team lists are usually quite dynamic. For example, the people who came up with the original idea may not be on the team. Key people may require recruiting, and the team may identify member "slots" for needed expertise, experience, or representation. Lists of member names may start on the back of envelopes, but they eventually become relatively formal, providing a roster of fundamental team identity.

A list of names offers an additional bit of basic information about the team: its size. However, membership boundaries may be less than exact, particularly in the beginning. Thus size is sometimes expressed as a range (for example, 8–10 people). Membership is often a moving target for virtual teams, becoming increasingly complex as the team grows.

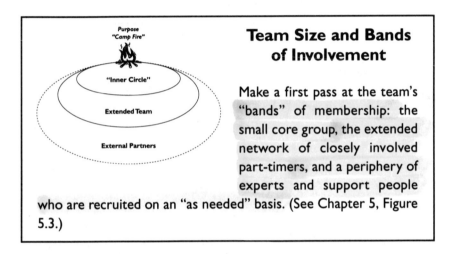

Team Size and Bands of Involvement

Make a first pass at the team's "bands" of membership: the small core group, the extended network of closely involved part-timers, and a periphery of experts and support people who are recruited on an "as needed" basis. (See Chapter 5, Figure 5.3.)

Make Connections

Team formation is above all a communication process. To reach people in the virtual world, you need to know their many addresses. Not incidentally, these addresses offer strong clues about people's current access to different media and their ability to operate in a distributed environment.

Contact and location information is central to the team directory—it is its catalog of boundaries and the means of crossing them.

Contact Information

Name/Organization	Medium	Address
...	...	...

Collect the many addresses that people and organizations have including office locations, snail mail (traditional postal) addresses, phone numbers (which may be numerous—office, home, car, cell, voicemail), fax numbers, e-mail accounts (perhaps several), Web page addresses, server names, and meeting places.

Do Assessments

Why, what, who, where, and *when.* By this point, you have asked the basic questions and have settled some initial team parameters. You probably know by now what kind of virtual team is coming to life.

Team Types

Spacetime	Organization	
	Same	Different
Same	Collocated	Collocated Cross-Organizational
Different	Distributed	Distributed Cross-Organizational

Look at your list of member organizations and where people are located to see the boundaries the team crosses. Use the chart to determine the virtual type. (See Chapter 2, Figure 2.3.) Consider how much time the team will spend together (synchronous) and how much asynchronous interaction it requires.

If a virtual team is distributed, but from the same organization, it is likely that communications and participation issues will dominate its development. A collocated but cross-organizational team is likely to experience difficulty with establishing common purpose and making decisions. Virtual teams that are both distributed and cross-organizational will experience both forms of stress and are most in need of new behaviors and support infrastructures.

With purpose, size, and timeframe in hand, you can also make some overall judgments about the complexity of the project, its chances of success, and its estimated costs. Look as well at the technologies that will affect the team's product or process. Note especially how much you can do online. Go digital wherever possible, as soon as possible, all other things being equal.

Now is the time to consider major changes! Encourage change early in the process and discourage it later.

This level of detail may be all you need. You ought not burden a relatively short and simple project with a few known players with unnecessary planning. If you can settle the basics in a few two-hour meetings and summarize them in a couple of pages, you need go no further in designing your virtual team. However, you still need to cycle through the essential elements described above several times. This guarantees that everyone involved will thoroughly understand the right mental models.

Always invest in beginnings.

Phase 2: Planning for Action

With your simple model of "who you are and what you are doing" in hand, you are ready for greater detail. Usually this second planning cycle involves more people, including both sponsors and a critical mass of individuals who are responsible for implementation and results. Participatory planning is a powerful way to achieve early virtual team alignment.

Two important outcomes of this phase are a responsibility matrix and a team handbook.

Purpose in Motion

Purpose is dynamic and complex, not static and simple. Virtual teams use their purpose to set themselves in motion and guide their daily work.

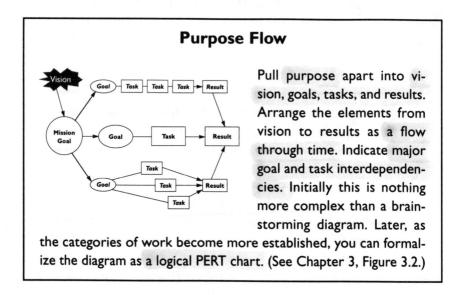

Purpose Flow

Pull purpose apart into vision, goals, tasks, and results. Arrange the elements from vision to results as a flow through time. Indicate major goal and task interdependencies. Initially this is nothing more complex than a brainstorming diagram. Later, as the categories of work become more established, you can formalize the diagram as a logical PERT chart. (See Chapter 3, Figure 3.2.)

As you cycle back to look at purpose for a second time, magnify it at different levels. Use a more inclusive frame of reference to set the statement of purpose in a broad context of vision, principles, and values. Then break the mission down to the next level of goals, tasks, and results. While it is useful to meet face-to-face for this part of the process, it is not mandatory. Often it is not even possible. However, you need to achieve the same results using interactive media, a clear process, and common information.

Agreed-upon tasks are the ways people spend their time in accord with the team's goals and results. Far too many teams are off and running with only a task list written on a flip chart and little idea of how the tasks connect to a larger purpose or specific results. For virtual teams, the list of tasks is the last part of purpose to talk about, not the first.[1]

Identify Process Elements

Work is a process of tasks accomplished, meetings held, decisions made, milestones met, and results produced.

Process Elements by Goal

Goal	Tasks	Results
...	...	...

Now is the time to detail the purpose. Check to see that each goal has at least one result and that each result has at least one task. Ensure that tasks clearly connect goals to results. Also identify key meetings, decisions, and milestones as you think through achieving the goals.

At the heart of a team's working model is its task set. Tasks are the execution center of the work plan to which members, leaders, resources, time, and other elements are attached. Return to goals, tasks, and results frequently and change them as new people and activities connect to them. Question whether you have divided the work intelligently and labeled it properly, and whether you need to develop new relationships. Until the end of Phase 3, these categories should remain dynamic.

All the actual activities and interactions involved in doing the work itself lie inside the "chunks" of work labeled as tasks.

Timing tasks generate a schedule that you can represent with that familiar icon of work plans, the horizontal time bars of a Gantt chart. Planning estimates turn into actual elapsed times as the team tracks itself

---

Task Timing

Task	Start	Finish
...	...	...

For each task, attach start and finish dates. If you know the over-all time frame, you can estimate specific deadlines. Use a flow diagram to evaluate task/time interdependencies. Refine your deadline projections as the project progresses based on ongoing actual information.

---

through the process. This basic project data allows the planning process to seamlessly become the management process.

Tasks may need no more than an ending date to indicate when specific actions need to occur in order to produce results. Indeed, people often resolve the trade-off between independence and interdependence by agreeing upon results and end dates. This balance between everyone working together, working in subteams, and working independently has to be flexible.

Clarify Responsibility

Tasks undergird the team's membership needs. A simple responsibility matrix captures the set of relationships between members and tasks.

One of the hidden dangers always lurking on the sidelines of virtual teams is the ethic that everyone needs to be involved in everything. You can avoid this recipe for disaster by clarifying just which tasks and decisions need everyone's input and which do not. The rich conversation about who needs to be involved in each task helps people to sort through and reduce anxiety about what is attainable. The exercise inevitably flushes out additional needs for expertise and representation, leading to new recruitment and perhaps a larger team. At the same time, people often reevaluate, cluster, or break out tasks further during this review.

Responsibility Matrix

	Member A	Member B	Member C
Task 1	x	x	x
Task 2		x	x
Task 3		x	

Identify "who needs be involved in what" through a dialogue, either in person, on the phone, or online, that examines each task. Whenever someone is involved in a particular task, make an x in that member's column. Some tasks require only one person. Others may call for everyone's involvement. Most tasks need a subset of the team as a whole. Use this chart as a way to explicitly divide the work, obtain the right participation, identify leadership, and track commitments.

Task and Process Leadership

For each task, designate one or more members as task drivers, specific members who are responsible for specific results.

Task Leadership

Tasks	Leadership
...	...

Identify who is responsible for what. Check that each task has at least one leader. Tasks lead to results, so every result will have at least one person responsible for it. Leadership may be singular or multiple, determined in the course of the task-by-task dialogue.

Virtual teams increase their overall leadership capability as they divide the work. By identifying task-based leadership, a group distributes its management burden in this team-defining dimension. A team may go into a planning session with one appointed leader and come out with everyone a leader. Leadership acts as a repository of trust within the group. The higher the level of trust, the less people will feel the need to be involved in everything.

Task leadership alone is not sufficient for team success.

Process Leadership

Name	Process Role
...	...

Process roles include team leadership as well as such functions as liaison, facilitation, knowledge development, agendas, and support. Be inventive when labeling process roles. Remember that this is frontier territory in the virtual world.

It is relatively easy to make task leadership explicit. Not so with process (social) leadership. Aside from overall team leadership, the roles required to develop and maintain the team process are hard enough to recognize and acknowledge in collocated situations.

Make Models Accessible

The explicit, visible models that a team creates are considerably easier to depict and access than the models that individuals hold inside their heads. As distributed work becomes more complex, the digital medium is the ideal way to make the team's process and product models come alive.

While product models, the topic of the team's work, will vary enormously, process models have an underlying similarity. The responsibility matrix is a powerful way to summarize purpose/people/links information that is key to every virtual team.

Computer-based tools can help you depict and update a relationship-oriented process model of virtual teams. The deployment charting method was pioneered by Dr. W. Edwards Deming and first extensively employed by Toyota in the 1960s.[2] TeamFlow[3] (Figure 8.1) is an artfully simple process mapping application designed for modeling and managing virtual team processes. At the heart of the deployment model is the responsibility matrix that reflects agreements about who is doing what.

Figure 8.1   Virtual Team Model and TeamFlow

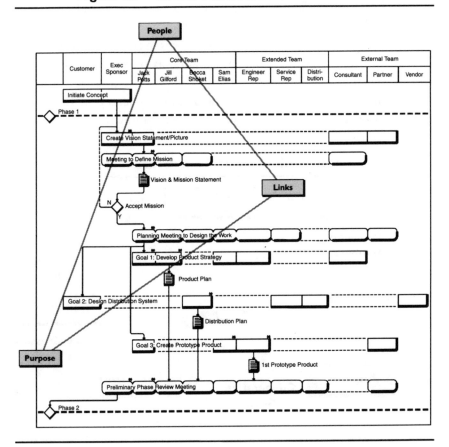

You can connect information that touches on all nine Virtual Team Principles to the process matrix and its two associated views—a time-oriented Gantt chart to track progress and an organization chart with a directory.

1. *Goals* serve as headings for groups of tasks and results.
2. *Tasks* are the set of linked boxes as a whole.
3. *Results* appear as document icons.
4. *Members,* individual and organizational, comprise the horizontal labels at the top of the matrix.
5. *Leadership* is indicated by tags on the tasks designating roles.
6. *Levels* are reflected in the hierarchy of the organizational chart and the levels of detail in the plans.
7. *Media* are accessed through addresses in the member directory and file names of other information associated with tasks and results.
8. *Interactions* happen over time and are indicated by the timing associated with each process element, summarized in a Gantt chart.

Figure 8.2   Virtual Team Pocket Tool

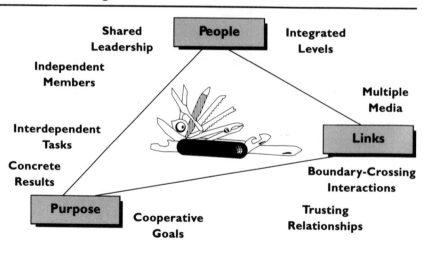

9. *Relationships* among members are represented by the linked boxes showing who is involved in each activity.

This information makes the principles concrete and useful. Remember the principles[4] as a tool set (Figure 8.2) to assess, plan, implement, and evaluate a team's work. Use them as a simple mental checklist or as a framework for creating a formal plan (as above).

Create Media Plan

The responsibility matrix and leadership roster indicate who needs to develop relationships with whom to complete different aspects of the work. Many virtual teams need or may greatly benefit from face-to-face time, particularly at the beginning to develop the plan and build trust. Since physical separation is the most common plight of virtual teams, it

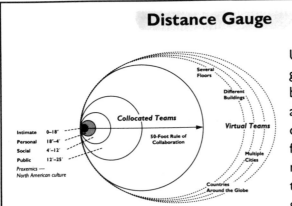

Distance Gauge

Use the distance gauge on a person-by-person basis to assess the likely frequency of face-to-face contact and the relative cost of the team getting together. The 50-foot rule suggests that people have to be very close together to gain the advantage of spontaneous interaction. Beyond that close range, cost and travel time increase with each greater distance. When people are face-to-face, take into account cultural differences in regard to personal proximity "comfort zones." (See Chapter 1, Figure 1.1.)

is important to evaluate the impact of distance in thinking about the team's communication.

Now that you have a feel for the work flow and time constraints, you are ready to develop a media plan. Consider your many options at this time.

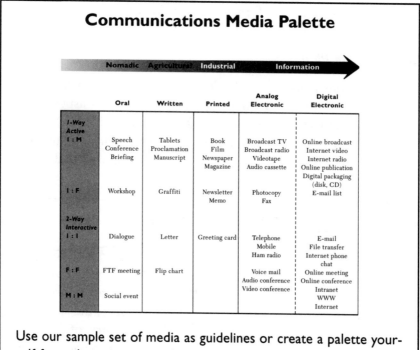

Communications Media Palette

Nomadic Agricultural Industrial   Information →

	Oral	Written	Printed	Analog Electronic	Digital Electronic
1-Way Active					
1 : M	Speech Conference Briefing	Tablets Proclamation Manuscript	Book Film Newspaper Magazine	Broadcast TV Broadcast radio Videotape Audio cassette	Online broadcast Internet video Internet radio Online publication Digital packaging (disk, CD)
1 : F	Workshop	Graffiti	Newsletter Memo	Photocopy Fax	E-mail list
2-Way Interactive					
1 : 1	Dialogue	Letter	Greeting card	Telephone Mobile Ham radio	E-mail File transfer Internet phone chat
F : F	FTF meeting	Flip chart		Voice mail Audio conference Video conference	Online meeting Online conference Intranet
M : M	Social event				WWW Internet

Use our sample set of media as guidelines or create a palette yourself from the systems you know about and have available. Choose from all eras of communications to best fit the work and process needs. (See Chapter 4, Figure 4.1.)

Virtual teams need to consider both product and process media. Product media relate to how you present the work you are doing. Process media relate to how you communicate in order to get the work done. First, will you deliver any results in a communication medium (for example, a document or presentation)? What work and results can you do in digital form so that they are accessible online? After you know what the work is and whom to involve, ask what media you need to carry the team itself and its work process of boundary-crossing interactions.

Media Plan

Media	Type	Interaction	Frequency	Location
...	...	...	...	...

Fill in the name of the communication medium you are using. Determine which of the five types it is: oral, written, print, analog (electronic), or digital (electronic). Designate whether it fosters one-way or two-way interaction. Note its frequency: periodic (for example, weekly, monthly), continuous (for example, e-mail), or variable and episodic (for example, special meetings). Indicate its location, physical or digital, as appropriate. (See Tetra Pak example, Chapter 5, Figure 5.1.)

A media plan may not amount to much if it simply mandates a face-to-face Monday morning meeting, a typical collocated team approach to staying connected. Virtual teams, however, require multiple media to meet a variety of product and process needs.

Look for the most appropriate media for meeting your needs, with obvious consideration of the limits of cost and availability. You may want a video conferencing system but find that it is currently too expensive for you. Within your constraints, experiment to find what works best. Then stretch your sights, particularly to find ways in which digital media can be useful to you. Digital media are ideal for virtual teams and will eventually become ubiquitous.

Virtual Team Handbook

By the end of Phase 2, you have accumulated considerable detail about your team and its work. Now is the time to pull it together into a team

Members/Media Matrix

	Member A	Member B	Member C
Medium 1	x	x	x
Medium 2	x		
Medium 3	x		x

In determining the communication plan, identify who has access to which media. Have people indicate their relative preferences for different media or their willingness to gain access to a medium they do not currently use.

handbook. A handbook can act as a shared "information place," a common resource for team members that is a tangible token of membership and a means for initiating new members. The physical handbook itself may be as small as a document or a file folder but it often comes packaged as a three-ring binder. The sections and content of a handbook are excellent models for setting up a digital workplace (see next section). It is best to have both a "carry-it-around" handbook of key information as well as a more extensive online version.

The virtual team handbook contains both process and product information:

1. Process information includes everything from purpose statements and team directories to budgets and instructions for using databases.
2. Product information, which always is unique to the team, includes everything from product requirements to marketing plans and the product itself.

Process Information

Organize process-related information for the handbook in three sections reflecting the purpose, people, and links aspects of the team.

Section 1: Purpose and Plan

Gather together all the purpose materials here including the statements of vision, mission, and goals, action plans with tasks and results, matricies of responsibilities, and decisions as they are made.

In the early part of a team's life, this section is very dynamic, changing and growing moment by moment. As the team's plan stabilizes, this section becomes a repository of that stability, serving as an internal management and navigation tool. Externally, the summarized purpose statement becomes the public face of the team answering the question, "What are you doing?"

Section 2: Team Directory

Review and update the contact information. Expand the list to include the new members whom you have identified by the more detailed look at the work. Broaden the information about people to include team roles and task responsibilities. Develop a format for entering organizational information by name.

A virtual team has external contacts as well as internal interactions. Team directories may include all the people and organizational contact information required in the course of its work. At the front of the directory section, post a team membership list to keep boundaries as clear as practical. You can represent complex teams in terms of bands of involvement—for example, the core team on one page, the extended team on the next, and all others such as occasional experts and reviewers in a third group.

Section 3: Calendar and Communications

Summarize all the relevant dates on a team calendar including result deadlines, task completion milestones, and scheduled events as well as holidays and other significant dates that impact timing. Capture the group's stream of consciousness, for example, by collections of meeting minutes, memos, and key group messages.

Getting calendars in sync is a major problem for virtual teams, along with the fact that published information always seems to be out of date. Online event lists, calendars, and scheduling systems address these problems and are essential parts of the virtual team's infrastructure. Store and organize all of the online communication for virtually instant recall and reuse, uniquely powerful features offered by digital media.

Product Information

The second part of the team handbook is organized according to task and results-related information. Constituting the bulk of the handbook as the work progresses, this section includes background documents, information related to work-in-progress, and drafts of results. Since even small groups generate a lot of information, use the binder as a selection device to keep key information pruned to the essential elements. Provide summaries, tables of contents, and pointers to other relevant documents and data wherever possible.

A well-organized, task-based plan with a responsibility matrix and common information gathered into a virtual team handbook can be very low tech and uncomplicated. This plan contains enough detail for alignment without being so detailed that it unnecessarily constrains people. For many virtual teams, this is enough.

Phase 3: Managing the Process

Teams that are particularly large, complex, long-lasting, or interconnected to other teams will need a third cycle of planning and a more comprehensive set of tools. Even if the level of detail you have achieved in the first two iterations satisfies you, review your task list against the eight types of activities discussed in this section.

One outcome might be to develop your team handbook on the Web. Also consider using the responsibility matrix as a group interface to common process information.

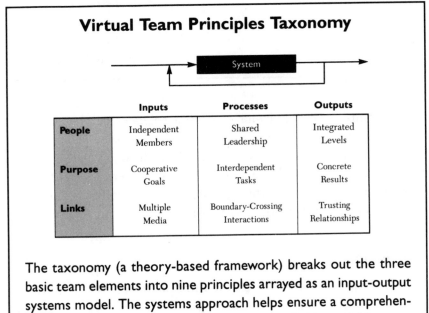

Virtual Team Principles Taxonomy

	Inputs	Processes	Outputs
People	Independent Members	Shared Leadership	Integrated Levels
Purpose	Cooperative Goals	Interdependent Tasks	Concrete Results
Links	Multiple Media	Boundary-Crossing Interactions	Trusting Relationships

The taxonomy (a theory-based framework) breaks out the three basic team elements into nine principles arrayed as an input-output systems model. The systems approach helps ensure a comprehensive look at team complexities. (See Chapter 2, Figure 2.5.)

Teams are truly varied, each as unique as a person. Yet, there are patterns of life and behavior that are common to all teams. The virtual team principles enable you to evaluate your own situation systematically. Adapt them to your unique circumstances.

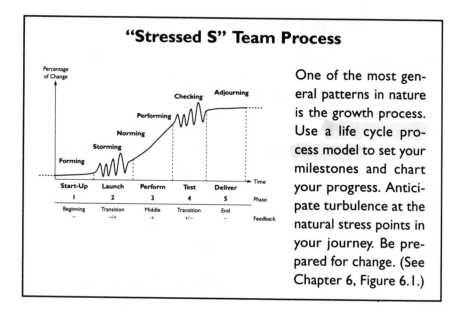

"Stressed S" Team Process

One of the most general patterns in nature is the growth process. Use a life cycle process model to set your milestones and chart your progress. Anticipate turbulence at the natural stress points in your journey. Be prepared for change. (See Chapter 6, Figure 6.1.)

By the third cycle, the planning process accelerates and the team's activities push into the performance phase. More detailed planning usually takes place in a more distributed mode. Individuals and subgroups develop many of the additional details while the team is apart. Then when the team comes together (physically or virtually), it integrates them to complete and legitimize the plan.

Type the Tasks

As with the previous two phases, the third phase begins with purpose. Here we introduce another tool to help us flesh out tasks, the defining quality of teams. If you can identify the type of task you are performing, you have at your disposal a large body of "know-how" about teams full of tips, tools, techniques, methods, and processes.

While the content of any particular virtual team's task is unique, the general form of the tasks is not. Just as you can categorize tasks as cooperative or competitive, independent or codependent, so you can categorize them according to the type of interactions they require. Joseph McGrath, a social psychologist at the University of Illinois, suggests four basic types of tasks, which he labels with verbs: generate, choose, negotiate, and execute. He arranges these types in a pie, the "McGrath Task Circumplex" (Figure 8.3).[5]

Imagine the process of building a house. Pounding nails and pulling wire—execution tasks—create the concrete result of a livable house. Putting the right nail in the right place at the right time involves many other activities. Owners and architects plan and create and plan again. Architects and engineers solve problems of load and leverage in design. Interior designers work with owners and architects to make decisions that depend on taste rather than fact. Owners, contractors, and other vendors negotiate over who does what for how much. Some work is put out for competitive bid.

Figure 8.3   McGrath Task Circumplex

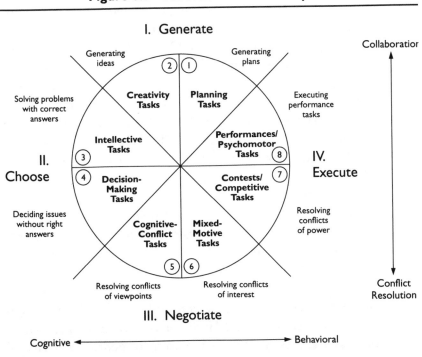

Note: Adapted from Groups: Interaction & Performance by McGrath, 1984, Upper Saddle River, NJ: Prentice-Hall, p. 61, Figure 5.1. Copyright © 1984. Adapted by permission of Prentice-Hall, Inc., Upper Saddle River, NJ.

All these activities lie behind the driven nail. Building a house appears to be the archetype of a project requiring a collocated team. However, the larger "building-a-house-from-beginning-to-end" process clearly involves a virtual team of many specialties, usually even a team-net, a virtual team of teams.

To open the right tool drawer, you must first appropriately identify the type of task you need to perform. Each of the types carries its own way of doing things:

- *Generating tasks*—which include (1) *planning tasks* to create plans and (2) *creativity tasks* to generate ideas—are the most collaborative ones. They involve a mix of thinking and doing activities. Usually, there is no "most correct" outcome to new plans and ideas. Planning is an iterative process of work design, while creativity involves divergent data gathering and brainstorming together with convergent idea integration.
- *Choosing tasks*—which include (3) *"intellective"* activities focused on solving problems with correct answers and (4) *decision-making tasks* where there are no right answers—are the most cognitive ones. Invariably they involve individual and group thinking. The intellective are the traditional science tasks where "truth wins" as judged by data and a jury of peers. Decision making under conditions without clear external standards, where most of human and social science lies, deals in probabilities, extrapolations, and estimates.
- *Negotiating tasks*—which include (5) *cognitive-conflict tasks* about clashing viewpoints and (6) *mixed-motive tasks*, which are about resolving conflicts of interest—are the most difficult group activities. Conflict resolution activities are the most competitive tasks and, like the generating tasks, they have few clearly correct answers. Groups need to be flexible and inventive around these issues. Here, success requires negotiating processes and skills. This is the arena of political activity, of tough choices that often require executive involvement.
- *Executing tasks*—which include (7) *contests* and competitions that are physical resolutions of conflicts of power, and

(8) *performances* and other psychomotor activities that require joint action—are the most behavioral ones. Executing includes a mix of individual and group activities. This is the domain of direct action, and these tasks dominate in the operational roles of an organization.

Taken as a sequence—generating, choosing, negotiating, and executing—the range of tasks models a team's natural life cycle. Many kinds of teams begin with planning (1), then move through a period of data gathering and creativity (2), followed by problem-solving (3) that data and accepted principles can settle. The team settles remaining problems through a succession of more competitive conflict-resolution processes, from issues without right answers (4) to differing viewpoints (5) and interests (6). They escalate unresolved issues of power (7) for settlement by the hierarchy. As tasks clarify and issues resolve, direct actions (8) produce concrete outcomes. And keep in mind:

- The code of eight task types is a language for steps in any sequence of work. For example, some teams begin with a planning task (1), move directly to resolving conflicts of viewpoints (5), then back to creativity (2), do another loop of planning (1), and then go directly to execution (8).
- The team's dominant task may type it as a whole. Some teams are set up for planning, others for problem-solving, and still others for decision-making.

The practical reason for determining which type of tasks you are doing is to identify appropriate tools, processes, skills, and competencies. Use of this system can also help extend your knowledge of virtual teams. By using a classification scheme based on the type of work, you can compare best practices, methods, and trends across many virtual teams.

Your Type of Tasks

As you take a third look at the work, consider the risks and rewards for the people involved. All goals are not naturally cooperative endeavors.

Teams are frequently set up to resolve conflicts, such as a cross-functional executive team charted to cut costs.

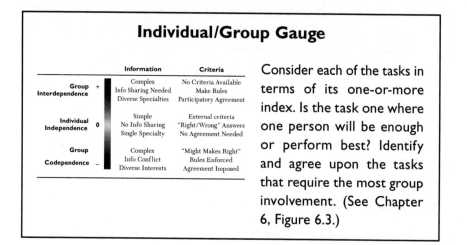

Cooperation/Competition Gauge

		Goals	Rewards
Group Interdependence	+	Cooperative	Big Win/Win
			Win/Win
Individual Independence	0	Independent	Individual Win or Lose
			Win/Lose
Group Codependence	–	Competitive	Big Lose/Lose

Consider the mission and each of the goals in terms of their cooperative index. Is it a win-win for the people working to achieve results? Or is it a win-lose competitive situation? Will winners and individual performers be rewarded separately? Will the team be rewarded as a whole? As part of the whole enterprise? (See Chapter 6, Figure 6.2.)

The cooperative character of goals colors the associated tasks and is the first step in typing the work. Second, consider tasks in terms of their suitability for individual or group effort. Teams typically involve both "me" and "we" tasks.

Individual/Group Gauge

		Information	Criteria
Group Interdependence	+	Complex Info Sharing Needed Diverse Specialties	No Criteria Available Make Rules Participatory Agreement
Individual Independence	0	Simple No Info Sharing Single Specialty	External criteria "Right/Wrong" Answers No Agreement Needed
Group Codependence	–	Complex Info Conflict Diverse Interests	"Might Makes Right" Rules Enforced Agreement Imposed

Consider each of the tasks in terms of its one-or-more index. Is the task one where one person will be enough or perform best? Identify and agree upon the tasks that require the most group involvement. (See Chapter 6, Figure 6.3.)

You can now assess each task with an array of indicators to better understand what it is and how to do it.

Task Factors

Task	Cooperative/ Competitive	Individual/ Group	McGrath Type	Tools/ Processes
...	...	...	...	...

Write down each task and then tag it in a variety of ways: Is the goal driving the task cooperative or competitive in nature? Will an individual or a group perform it best? Is it a generating, choosing, negotiating, or executing task? What specific processes and tools are best for each task type?

Many tasks require specific media to support their process and interim results.

Media Characteristics Chart

	Oral	Written	Printed	Analog Electronic	Digital Electronic
Interaction					
Space	Collocated	Distributed	Distributed	Distributed	Distributed
Time	Synchronous	Async	Async	Sync/Async	Sync/Async
Size	Small	Small	Mass	Unlimited	Unlimited
Speed					
Produce	Speaking	Writing	Write and Print	Real-time	Variable
Deliver	Sound	Transport	Transport	Electronic	Electronic
Receive	Hearing	Reading	Reading	Real-time	Variable
Delay	None	Some	Lots	None	None
Memory					
Store	None	Integral	Integral	Integral	Integral
Recall	None	Limited	Limited	Limited	Integral
Modify	None	Limited	Limited	Limited	Unlimited
Reprocess	Separate	Separate	Separate	Separate	Integral

Evaluate tasks for interactivity. What challenge does each task pose in crossing space and time? How much speed do you need and how much memory? Use the chart to identify the right media for the job. (See Chapter 4, Figure 4.2.)

Make judgments about the appropriateness of media depending on the task type. For example, media that are good for creativity are not necessarily best for negotiation.

Level with People

Good virtual team models provide multi-level views of the people involved in order to navigate the organizational space. The most powerful single conceptual tool people can use to develop and communicate meaningful mental models is the concept of levels (i.e., hierarchy in the scientific sense).

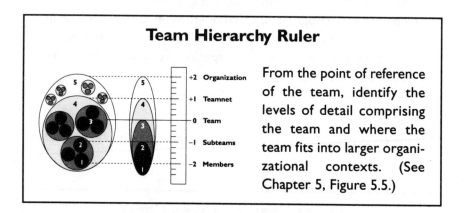

Team Hierarchy Ruler

+2 Organization
+1 Teamnet
0 Team
−1 Subteams
−2 Members

From the point of reference of the team, identify the levels of detail comprising the team and where the team fits into larger organizational contexts. (See Chapter 5, Figure 5.5.)

Hierarchical diagrams represent both authority relationships and the logic an organization has developed for its work. Virtual team directories should contain basic positional information that indicates who reports to whom (especially important in matrix organizations). Try rendering your many-tiered organization in the style of Eastman Chemical Company's Pizza Chart.

Finding the right people at the right time for a task is a challenge. Ideally, corporatewide Yellow Pages offer access to people through work experience, skills and competencies, educational records, other formal qualifications, and personal information that people choose to make available.

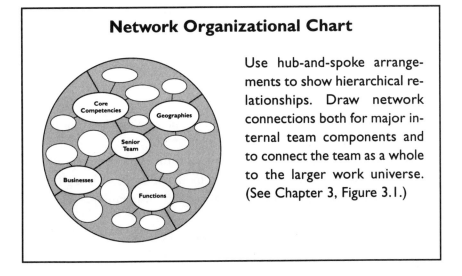

Network Organizational Chart

Core Competencies

Geographies

Senior Team

Businesses

Functions

Use hub-and-spoke arrangements to show hierarchical relationships. Draw network connections both for major internal team components and to connect the team as a whole to the larger work universe. (See Chapter 3, Figure 3.1.)

Virtual Team Web Book

In a few years, virtual team *Web* books will be almost as common as manual team *hand*books are now.

The World Wide Web stands out as the defining tool of the emerging world of virtual work. This is the portal to what is still the cyber frontier of the Internet and the even vaster matrix of all interconnected computers. In virtual team applications, using the Web means creating intranets.

An almost ideal medium for shared mental models, the Web combines anytime-anywhere availability with an unrivaled flexibility of expression that potentially embrace all media. At the hub of the Web is the *link*, not even a proper technology but a pure package of information, an address tag—a Uniform Resource Locator (URL).

On the Web, you can link any information to any other information anywhere. Most remarkably, you can put these links into context, however consciously or unconsciously constructed. A team's Web site is a place to clearly and explicitly grow its common information. Smart teams thereby become more conscious.

- Virtual team Web sites can be as simple as they need to be. A home page with a name banner and clickable references to current events, purpose, people, and communications may be all that is necessary.
- The next step is to create an online end-product model—a picture, description, and/or diagram of the results showing major components. This can act as a common navigation tool to the team's product.
- The third step is to use the Web with its interactive power to create a visible dynamic process model of the team using the information developed in the three cycles of planning.

The responsibility matrix created by a team can be a powerful *group* interface, its navigation tool for process information. All the elements can be hot, clickable URLs to associated information. For example, click on a name and go to a directory entry, click on a task and go to its description, or click on a link to learn what someone is doing on a task.

The Web approach to team intelligence is completely scaleable in size. Teams may use personal home pages (like those offered by many Internet Service Providers) or they may plug into global internal networks of intranet sites, as at Sun Microsystems, where they leverage vast resources. Modeling tools can combine product and process views at any level of complexity.

Remember that there are no rules for what you can and cannot include. Regardless of what information you provide, whether you keep it mostly in hard copy or move immediately to a Web-based resource, your virtual team handbook will be a tremendous source of information and even pride for your team.

Invest in a handbook or Web book and your virtual team will reap the benefits of being smart, nimble, and fast.

CHAPTER 9

VIRTUAL VALUES
Generating Social Capital

The world's largest center for diamond research is located in an unlikely spot: State College, Pennsylvania. There, in the early 1980s, scientists in the Materials Research Laboratory at Pennsylvania State University discovered that they could make industrial diamonds at atmospheric pressure. This finding proved so consequential to the industry that 30 competitors immediately joined the Diamond and Related Materials Consortium to explore the implications for their businesses.

"In the pre-competitive stage, it's pretty easy to do," says Rustum Roy, the physical chemist who runs the lab. "We're the leading research group, we have the equipment, and we have the people. So we say to the members, 'We're a networking center for you guys. When you come here, you learn from Penn State and each other. You'll say stuff here that you'd never say in a public meeting.'"

"Suppliers of equipment become partners of users through the glue of the lab," Roy says. Thus many of the consortium members have become customers of one Boston firm that makes diamond coating instruments. "This is what I mean by networking. What appears to be a competitive situation ends up as a complementary one."

When Roy notes that consortium members say things in the lab that they would never say in public, he is not talking about collusion. Rather he refers to the trust that develops among the members of this virtual team. As trust grows, people confide in one another more and more. They mutually learn from the give and take.

Trust in Teams

Virtual teams respond to the need for quicker, smarter, more flexible work groups in a sea of change. Indeed, these teams are highly adaptive social organizations that can cope with tumultuous complexity—like working with your arch competitors on projects that will benefit you both *and* allow you to compete better against each other. By learning from one another, each of the competitors becomes more skilled at what it does best.

Consider EBC Industries in Erie, Pennsylvania. It is one of small business's best examples of crossing competitive boundaries in the United States. Suffering from annual losses of $200,000 by the mid-1980s, the company's CEO Harry Brown turned to his competitors in the small metal parts manufacturing industry to solve his business problems. Working together, some 50 firms in and around Erie have teamed up on projects that none can do alone. When they finish the projects, they then return to competing for business that they can carry out by themselves. This alternation between competition and cooperation has proved profitable for the firms involved. Brown's revenues have quadrupled, employment is up, and profits have replaced losses.[1]

How can competitors work together? They face even more obstacles to trust than plague virtual teams from the same organization. Yet, boundary-crossing teams overall need more trust than do collocated teams. Without daily face-to-face cues, it is at once both harder to attain and easier to lose. Mistrust slips in between the slender lines of long-distance communication stripped of the nuances of in-person interaction.

Trust is the elixir of group life. Broadly speaking, trust is the belief or confidence in a person or organization's integrity, fairness, and reliability. This "matter of faith" comes from past experience, however brief or extensive. The importance of trust cuts across a team's life cycle:

- A new team requires trust in order to get started.
- Trust is the all-purpose grease for the ongoing hard work of the team.
- When it ends, a team leaves a legacy of trust (or lack thereof) to the organizational environment from which it came.

A virtual team must pay special attention to building trust at each stage of its development. All other things being equal, the benefits of high trust in an organization are self-evident: Teams with higher levels of trust coalesce more easily, organize their work more quickly, and manage themselves better. Lower levels of trust make it much more difficult to generate and sustain successful cross-organizational, cross-distance groups.

Trust has always been important for groups. In the work-a-day world of the Industrial Age, it was more a "nice to have" quality than a "need to have" one.

In the networks and virtual teams of the Information Age, trust is a "need to have" quality in productive relationships.

Beyond Integrity

However competitive the market, the road to profit runs through the by-ways of cooperation. Businesses and the teams that comprise them function because people work together. "Trust is mandatory for optimization of a system," wrote W. Edwards Deming, the inspiring founder of the quality movement. "Without trust, there cannot be cooperation between people, teams, departments, divisions. Without trust, each component will protect its own immediate interests to its own long-term detriment, and to the detriment of the entire system."[2]

Few organizations tolerate lying, cheating, and stealing. We all know the basic moral and practical costs of dishonesty. More subtle are the tokens of trust and mistrust that people convey through competence, rewards, and information. Each of the three virtual team elements—people, purpose, and links—offers a source of trust. Each also holds a potential for mistrust that goes beyond honesty and integrity.[3]

Trusting People

Of fundamental importance is trust in people and their competence. Task-oriented teams need more than trust in a person's integrity. If we do not trust people's competence, then we will not rely on them or the results of their work.

People demonstrate competence over time. Consequently proficiency can take longer to establish in virtual reality than it does face-to-face. Because we are just learning the skills of presenting information online, we are still in a period of skeptical acceptance of what others have to say. A person whose words read well on the screen may or may not appear to be knowledgeable in person. Likewise, proficiency can be more difficult to verify at a distance. If you can drop by someone's office, see first-hand examples of prior work, and talk with other colleagues, you can more easily evaluate their proficiency.

Occasional online interaction is just one step up from reading someone's resume. What may read well on paper may not translate into knowing someone in person. Reputation, recommendations, and resumes loom larger when people must establish relationships quickly through narrow channels. Likewise, online proficiency is easier to demonstrate in organizational cultures that support it, such as Buckman Laboratories (see Chapter 2), where the highly technical work of the company depends upon the virtual exchange of expertise.

"Trust comes from performance," says Lee Sproull, professor of management at Boston University, who has been following the development of trust and relationships in online environments for many years. "If I see this person is going to do a first-rate job with the information I provide, that [s]he won't undercut it, won't embarrass me, then I'm more likely to trust [that person]."[4]

Core-R.O.I., which specializes in developing labor-management partnerships, change management, and organization redesign, has operated as a virtual team since its inception in 1982. Its members are in New Mexico, Texas, Washington, DC, Iowa, and North Carolina. George

Gates, a consultant and partner, reports that the group intentionally meets at least four times a year. Initially, the group devoted most of its face-to-face time to reviewing the financial aspects of their business. Today, they spend most of their time talking about their practices and what they are learning. "We get together just to get together," Gates says. "You can use as much whiz bang technology as you want, sending parts of your head around on the Internet but you can't send parts of your heart. We all know how to type little smiley figures on the end of our sentences, but great, what does that really say?"[5] Gates' point is that by coming together, the group renews its basic trust and belief in one another, and reinforces the values that have held them together for 15 years. Icons on a screen do not a relationship make.

Trusting Purposes

The second way that people generate trust is by their commitment to a unifying purpose with shared rewards. Conversely, nothing provokes mistrust faster than a mismatch between a team's goals and the system that rewards it. Many companies ask people to work toward cooperative goals then evaluate and reward them on the basis of their individual performances. This often arouses suspicion and provokes people to act competitively. People have highly developed fairness detectors, particularly when their employers do not recognize or, worse, violate the relationship between contribution and reward.

A large pharmaceutical firm commissioned a cross-organizational virtual team to study cost reduction but the steering committee provided very little direction. "All they said they wanted were 'deliverables,'" reported the team leader. "It was very frustrating." With no objective measures of their success and no reward in sight for a job well done, team members were reluctant to give their unbridled support to the project.

Trusting Links

Third, people must trust information and their information channels. Because virtual teams are information-intensive, they rely heavily on the quality, quantity, and availability of information, making it a prime source of trust and mistrust. People expect to have what they believe is the best-available information to do their job. Organizations must keep some information private, such as salary and other personnel matters. Within the common space, however, partial, incorrect, misleading, and late sources of information are all potentiators of mistrust.

Open-book management[6] that advocates providing essential information to everyone in the organization is one way to contribute trust to the environment. If the company's financial information is available to people at all levels of the firm, then people regardless of their position are likely to feel a greater sense of trust among one another. By removing the privilege of private access to information, such companies elevate everyone's feeling of being on the inside.

Even when a company has the best of intentions, it can inadvertently create mistrust by not releasing enough information. A publishing firm with several thousand employees decided to hold an all-company meeting, the first of its kind in its history. Its purpose was to build greater communication channels among the ranks of the organization. For weeks before the meeting, rumors abounded that the company was being sold and that massive layoffs loomed. A simple memo from senior management prior to the event addressing the rumors directly could have saved the company hundreds of hours of time wasted on false speculation.

Trust is part of that difficult-to-grasp, nonmaterial world of relationships. Yet, relationships are increasingly being recognized as having true economic value. Indeed, relationships store a new form of productive capital.

Social Capital

Teamwork, whether collocated or virtual, generates a double bottom-line:

- Task success, the value of the results; and
- Social success, the value of the relationships.

Management invariably evaluates teams on the basis of their performance goals and the quality of their outcomes. Management also needs to evaluate teams on the quality of their interactions and the enduring relationships among members. Why does this matter? It matters because every small group leaves a legacy in the larger organization. It either adds to or depletes the existing stock of "relationship resources." The legacy of relationship success accrues as *social capital*.

All organizations have relationships based on past experience, ongoing interactions, and expectations for the future. Thus, all organizations, large and small, have some social capital that is continuously growing and diminishing, a hidden source of wealth or a deficit that may presage a disastrous weakness.

What Is Social Capital?

Social capital is "the structure of relations between and among actors,"[7] individual or organizational.

It is easy to demonstrate the value of social capital. Can you recall a friendship or professional relationship that you established in one team or small group that later proved to be a valuable connection in another context? Can you remember deciding to do business with an external partner, vendor, or customer because of its pre-existing organizational reputation?

Imagine the potential in your organization for better, faster, smarter relationships based on a rich network of pre-established lines of trust. Each strong relationship has a multiplier effect built into it: The "friends of friends of friends" are potentially accessible through social networks of trust. The old adage, "it's not what you know but who you know," portrays the colloquial acknowledgment of social capital.

The negative side of social capital also pertains. Do you recall cautioning others about people whom you came to distrust as a result of working with them? Did a team experience leave a "bad taste in your mouth" that affected other situations or opportunities? Have you seen previously good relationships between people or companies strained or snapped to the detriment of both? A bad experience also has a multiplier effect. People pass along news of mistrust, diminishing the capacity for collaboration within and between organizations.

It is better not to team at all than to team badly.

Social capital is a seminal idea that has been growing at the intersection of economics and sociology since its 1988 introduction in a paper by University of Chicago sociologist James Coleman.[8] For the most part, the idea flies below the radar of public consciousness but one mass media peek came in a 1996 ABC *World News Tonight* segment. It reported on Harvard government professor Robert Putnam's disturbing thesis that social capital is dangerously eroding in the U.S. society as a whole. Putnam illustrates his point by many measures of declining participation in civic and social events. Among them is the telling observation that while more Americans are bowling than ever before, they are "bowling alone" rather than in once-popular bowling leagues.[9] His influential 1993 book *Making Democracy Work* details how stocks of social capital a thousand years old were the best predictors of governmental and economic success and failure among Italian provinces in the 1970s and 1980s.[10]

How to Create Social Capital

It is easier to form, launch, and sustain virtual teams in an environment rich in "the features of social organization . . . that facilitate coordination and cooperation for mutual benefit," namely:

- Trust;
- Norms of reciprocity; and
- Dense social networks.

Putnam and his team of researchers identified these three factors as components of social capital. To work with people you rarely or never meet, you need some basis to believe in their expertise and trustworthiness. Clearly, a norm that supports dishonesty in some relationships rubs off on other relationships as the level of suspicion rises. The fragile sphere of virtual relationships requires a much higher level of trust than do conventional hierarchically controlled settings. Top-down control can mandate people to work together whether they want to or not. Virtual teams have only their shared trust in one another as their guarantee for the success of their joint work.

When Buckman Laboratories (see Chapter 2) began to expand into global markets, questions of integrity immediately arose. Among the concerns that employees raised was whether to pay bribes. The issue prompted the development of a code of ethics that has become central to the Buckman community. Among the 10 points in the code is this one: "That we must use the highest ethics to guide our business dealings to ensure that we are always proud to be a part of Buckman Laboratories."

The larger organizations that house good teams almost always have strongly expressed values embodied in codes, philosophies, and principles. They invariably include trust along with integrity, teamwork, and a commitment to the value of the individual. "The Eastman Way," a pillar of Eastman Chemical Company's corporate culture, declares, "Eastman people are the key to success. We must treat each other fairly and with respect, based on values and principles: honesty and integrity, fairness, trust, teamwork, diversity, employee well-being, citizenship, winning attitude."

The norm of reciprocity—you do something for me and I will do something for you—recognizes that a favor received will somehow be repaid in the future. The oft-used phrase, "I owe you one," speaks precisely to the value of a reciprocal relationship. Business is awash in these sorts of "owe-sies": People either barter directly—I will give you this piece of business if you give me that—or they bank obligations for the future. "We will all benefit later by working together now" underscores a belief in deferred gratification.

Dense social networks, Putnam's third component of social capital, are a hallmark of healthy communities and businesses. According to his research, the more involvements people have in community life, the stronger the economies of their regions. The same idea applies to business. The more activities that people engage in together, the greater their commitment to one another. Company picnics that include employees' families, online chat rooms where people can talk about their hobbies, and corporate support for community involvement all build social capital inside the company as well as outside.

Social capital is both an enabler and an outcome of good teaming.

Scaling Up

The idea that relationships of trust and cooperation can have productive benefits has sparked a revolution in the field of economic development. Social wealth, valuable in the business world, offers a powerful new development resource for people with limited human and physical capital.

At the United Nations Development Programme (UNDP), the decades-old vision of large-scale development projects has shifted to a new vision of "sustainable human development." The core of UNDP's strategy is to build social capital. Its mission is "the enlargement of people's choices and capabilities through the formation of social capital so as to meet as equitably as possible the needs of current generations without compromising the needs of future ones."[11] Social capital "places not just the human being at the center but, above all, the *relations* among human beings . . . because they constitute the basis on which moral communities are built. Human capital seeks to improve the ability of an individual to make decisions; social capital seeks to improve the ability of a collectivity to make decisions."

The UNDP represents a large-scale application of relationship wealth, literally to whole continents. One example of UNDP's country-spanning effort to create social capital is the African Management Development Network, where the purpose is to strengthen management capacity in both public and private sectors throughout Africa.

Far from UNDP outposts around the world, in Silicon Valley, California, social capital has been rapidly accumulating thanks to the pioneering efforts of Joint Venture Silicon Valley Network (JVSVN). Since

1988, leaders in business, government, education, and nonprofit organizations have been working together to address the overall problems of the Silicon Valley community, spawning projects that often make headlines. NetDay, for example, the effort to wire public school classrooms in California, which became a national effort, was the brainchild of Smart Valley,[12] one famous offspring of JVSVN.[13]

In her ground-breaking book, *Regional Advantage*,[14] AnnaLee Saxenian describes the culture of Silicon Valley as one that promotes collaboration across business and sectoral lines. She contrasts this "social capital building" environment with that of Boston's Route 128 region. There leaving one company to go to a competitor can be regarded as an act of heresy. From an economic standpoint, Saxenian observes that the recession of the late 1980s quickly reversed in Silicon Valley while the Route 128 region was still suffering well into the 1990s.[15]

JVSVN's work has inspired similar efforts in other communities that are profiled in *Grassroots Leaders in the New Economy: How Civic Entrepreneurs Are Building Prosperous Communities*. Such attempts to consciously build social capital are often the work of individual business people like Harry Brown of EBC Industries. Such entrepreneurs look beyond the traditional needs of their businesses—markets, employees, and funding—to the larger environment that makes it possible to maintain and capitalize on those resources. They recognize that unless there are highly trained people with the right skills coming out of universities, the local labor pool will top out and growth in the knowledge-based economy will stifle. Issues like this concern civic entrepreneurs and their colleagues in regional economic development collaboratives.[16]

Starting Small

Social capital affects every level of human organization and society:

- Every virtual team member whom we interviewed—whether in a group of 5 or 50—affirmed the singular significance of trust within and between teams;

- The half-decade-old multidivision team developing Hewlett-Packard's worldwide distributed product information management system (PIM System) gains long-term value from trust and relationships;
- Companies like Eastman and Buckman Labs show the value of trust for growth and profit at the enterprise level;
- Saxenian draws conclusions about the value of relationship riches within states or regions in her comparison of high-tech industries in California's Silicon Valley and Massachusetts' Route 128 high-technology business beltway;
- Putnam and his colleagues documented the impact of stored trust at the country level through the example of Italy; and
- The UNDP illustrates the value of relationship capital that reaches across countries.

Great efforts begin with small ones. Small groups, constituting the "cells" of all larger organizations, fundamentally comprise human societies at all scales in all sectors.[17] Trust originates in teams as well as in other small groups—families, friendships, and myriad formal and informal associations based on shared interests and common concerns.

Even with its vast global purview, the UNDP recognizes that the formation of social capital starts small. "It gives the edge to small scale [as] it is in such contexts that social capital is most effectively formed."[18]

To grow trust, small is beautiful.

For goal-oriented, task-based business organizations, *teams* are the "cells." At work, we interact with others for largely task-oriented purposes. We cannot avoid teaming. We can only team well or badly, consciously or unconsciously.

Thus, we will accrue or deplete our corporate social capital with every small group in the organization, whether we consciously acknowledge the value of relationships or not.

Capital over the Ages

Virtual teams include features from all the eras of human organization. Success demands both ancient skills of small group interaction and co-operation *and* emerging skills of communications and knowledge development. Virtual work brings human beings on a great return to their earliest roots—as small groups that cross "family" lines.

The 21st-century return to our many-millions-year-old-roots carries a quiver of new collaborative tools of awesome power. Social capital is an old form of wealth, albeit largely unacknowledged. Suddenly, however, we have new ways to create and magnify it outside the confines of physical spacetime limits. With the ability to reach across great distances without having to travel them physically, we are able to build communities of high trust that circle the globe.

Unlike human and physical capital, individuals cannot possess social capital. It lies in the web of relationships among us and mingles with other means of generating wealth.

The Evolution of Capital

Capital—physical, human, or social—facilitates productive activity. Forms of capital have accumulated over the great eras of human civilization:

- *Human capital* is a concept developed in the 1960s as a way to describe the value of the people part of the work equation, the skills and knowledge of individuals. The oldest form of capital, reaching back to the earliest societies, it is rooted in people's ability to survive in the world around them. As environmental challenges change, so do the attributes of survival and success that make up human capital. Thus, new knowledge-based skills that people need in the Information Age replace many of those required in the Industrial, Agricultural, and Nomadic eras.
- *Social capital* is the complement of human capital, reflecting the community skills that have co-evolved with individual skills. People working together generate webs of social capital. Hunters

and gatherers compensated for resource scarcity by pooling their communal smarts. Today, people can form social capital abundantly and omnipresently, no longer constrained by space and time.

- *Land capital* harks back to the economic basis of the Agricultural Era. With farming and herding, people used land in an entirely new way to provide a relatively predictable food supply. In domesticating aspects of nature, human beings took a dramatic leap in scale and civilization. In humanity's earliest period, people prized but did not individually own the land and its bounties. In the next era, the hierarch, whether high priest or warrior, possessed the land. Herein lie the origins of ownership.

- *Machine capital* became the great engine of economic growth in the Industrial Era. Technology rolled on with the laws of motion, remaking the world from hand tools to locomotives. People generated new fortunes with productive machinery, but fields remained fertile. Land did not cease to have value as machines became dominant. Even at the end of the 20th century, people still perceive technology as the most potent force in economic growth.

- *Knowledge capital,* as an organizational source of productive capacity, resides in all the shared repositories of information and learning. Digital cyberspace offers a vast new domain for this old source of wealth that is newly powerful and available in historically novel ways. At the millennium's turn, information products and services spur growth and hope for an expanding economic future for all.

The recognition of knowledge capital and its value are at the competitive cutting edge of the global marketplace.

Shared knowledge will be the dominant productive source of 21st-century economics with consequences we cannot now even imagine.

Accumulating Capital Virtually

Virtual teams possess human capital *in* their members, and social capital *between* their members. They utilize physical capital that is *outside* people through their meeting facilities and communication infrastructures.

Cross-boundary groups also generate knowledge capital that exists in all three forms: *inside* people in memory and internal cognitive models; *outside* people in commonly accessible information such as databases; and *between* people as they connect parts and pools of knowledge together and develop enduring understandings.

All virtual teams, whatever their specific tasks, can increase human, social, and knowledge capital in particular. By working with more people in more places, human capital increases as individuals meet new challenges and acquire new competencies. Social capital accumulates as virtual team members vastly expand the number and diversity of their relationships. Because of their physical separation, virtual teams have an obligation to make knowledge capital explicit and accessible.

By stretching the bounds of human capability, virtual teams offer value far beyond their immediate functions: They stretch the reach of the social capital they generate outside their immediate physical locales. Although many of their elements have ancient roots, today's virtual teams look out over vistas of virtual places never before seen by human eyes.

The new frontier is not far away; it is everywhere.

At the Frontier

Cyber frontier: We and other writers have perhaps too often used the frontier analogy with respect to cyberspace. Thus, it is worth listening to someone who has been to "the end of the earth" for a reminder of just how really appropriate it is.

John Lawrence, who organized a World Wide Web site for the 1995 United Nations Fourth World Conference on Women in Beijing (see Chapter 7), is a fascinating character among those involved in the creation of electronic places. At one time a geological explorer for the New Zealand Antarctic Research Programme (Lawrence Peaks, which is part of the Transantarctic Range in Victoria Land, Antarctica, is named for him), he says today: "One simply trades one form of frontier for another. I know the feeling of stepping out onto land that no human in recorded history has stepped on. It was a feeling very similar to what cyber people are feeling now as they go out into this peculiar virtual world."

Lawrence continues, "There's an adrenaline rush as one goes over new surfaces, seeing completely new vistas that have never before been seen by the human eye. It's incredibly exciting and each person has his and her own way of coding all that. But this is different and more intriguing because explorers have gone out into new territory in physical space for hundreds of years. That particular adrenaline rush has been described for generations. But this new one has barely been described for a generation and that's a rush in itself."

Of course, cyber explorers can be anywhere.

Protecting Prairies with a Screwdriver

Sitting in Fergus Falls, Minnesota, population 12,000, Peter Buesseler is a pioneer in the use of virtual teams. He is a key node in the Great Plains Partnership (GPP), the initiative of 13 western states, three Canadian provinces, two Mexican states, numerous federal and local agencies, American Indian tribes, environmental and agricultural organizations, businesses, and landowners concerned with the viability of the Great Plains.[19] He is also "Webmaster"[20] of the GPP World Wide Web site.

"How am I protecting prairies while I'm going around with a screwdriver in my pocket?" asks Buesseler, Minnesota's State Prairie Biologist, and friend to many Minnesotans who are trying to get online. "We're in a rural part of the country here and e-mail is not much available. I'm often involved in helping people I need to work with find out what kind of access is available to them. I talk to the telephone companies for them, and then take my screwdriver with me to their offices or homes to attach their modems."

Ten years ago, Buesseler could not even type. Since then he has turned himself into "a little techie," he says, in order to be able to reach the people he needs to work with. "It's a lot easier for me to do it than for them to wait three months if they made the same request from their data centers. It builds a relationship that is not as structured. We can ask each other for things that we might not think to ask each other. It's a barn-building type of arrangement which gets at the core of my work."

We conducted our interview with Buesseler, along with two of his colleagues, Brian Stenquist, a senior planner in Minnesota's Division of Fish and Wildlife, and Susen Fagrelius, who consults to the state's Department of Natural Resources, via conference call, naturally, as the three of them were 300 miles apart and we were in Boston. At one point Buesseler said, "I am sitting here mentally doodling spider webs which are held together and anchored at key strategic points all the way around. But the material that it takes to hold them together is pretty minor. It's both very delicate and incredibly strong at the same time. A spider can walk across it but other insects that try to walk across get entangled."

Buesseler clearly draws the analogy to the network that each virtual team spins—at once fragile but strong, unique yet constantly changing, dependent on its environment that it reconfigures to its best advantage.

"You can tell the species of a spider by the pattern of its web. Each is different and no spider will ever make the same web twice. It's always dependent on the environment. Is it using a twig or a doorway? In the morning, it is beautiful and glistening, but it is in constant need of repair and demands a lot of upkeep. Its design is always contextual, always aware of its environment and drawing its elements together." The same is true for virtual teams.

Changing the World

Visions of the future are replete with new technologies, mostly extensions of the current state-of-the-art, and their impact on business and everyday life. The most profound change in the next few decades, however, may well be organizational as a trend thousands of years old suddenly reverses.

Society established the "bigger is better" trend in organizations long ago. At the dawn of the Agricultural Era, the average size of camps suddenly grew from a nomadic 20 to a farming community of 200. "Bigger" has had a largely uninterrupted run for 12,000 years until the end of the 20th century. In a comparative nanosecond of evolutionary time, centralization has reached global limits. Expanding information access has rendered hierarchical control both difficult and unnecessary.

The 21st-century trend will be that "smarter is better." Smarter teams and small groups of all types are the cells of more intelligent organizations of every size and sector, from family to humanity as a whole.

Imagine kicking our ability to team up a level. Improving our collective capabilities of teaming improves everyone's ability to solve their own problems. With more effective working groups we also can take up challenges with others that are currently impossible to achieve. This is true whether the scale is a few entrepreneurs who form a flexible business network or a group of countries who organize to meet the challenge of global warming.

Communication technologies and computer networks—in particular the Internet—are underwriting this moment of pregnant potential. Astonishingly enough, the possibility of a leap in social capability will bring individuals and small group relationships back to center stage.

Americans enshrine their personal freedom and independence in their Bill of Rights, the first right being that of free speech. In its 1996 opinion extending First Amendment rights to cyberspace, a three-judge U.S. federal panel wrote:

> The Internet may be fairly regarded as a never-ending worldwide conversation. The government may not, through the [Communications Decency Act], interrupt that conversation. As the most participatory form of mass speech yet developed, the Internet deserves the highest protection from governmental intrusion.[21]

As more people become interconnected through computers, our human capacity for both independence and interdependence increases. We are creating new environments where both competition and cooperation thrive. The global Internet is a way to foster innumerable combinations of groups of every size, while also sponsoring mass individuality and participation. Cyberspace is a vast new marketplace, containing both places of commerce and an already rich social life reflected in countless conversations.

We are only just beginning to learn about virtual teams and the world(s) they populate. The people who spoke to us for this book—CEOs, team leaders, and public servants alike—are harbingers of the world of work of the future—crossing space, time, and organizations.

In time, virtual teams will become the "natural way," nothing special. Virtual teams and networks—effective, value-based, swiftly reconfiguring, high performance, cost sensitive, and decentralized—will profoundly reshape our shared world. As members of many virtual groups, we will all contribute to these ephemeral webs of relationships that weave together our future.

AFTERWORD

www.netage.com

Want more information? Check out our World Wide Web site at http://www.netage.com. It is amazing how suddenly prevalent this new form of address is.

Our place on the web reflects our vision of a better working world through better ways to work together. On the site you will find information about:

- Virtual teams, teamnets, and networks;
- The Networking Institute, our small business;
- Our own experience as speakers and consultants; and
- Our network of partner organizations.

Soon online Web books will become a common complement to today's three-ring team handbooks. If you are in this vanguard, you will find material here that you can reference in your own Web book, making the link as close as a click away.

More information is also available to you through the traditional publishing medium. If you find this book on virtual teams insightful, useful, or otherwise enjoyable, you will find even more in our companion books on teamnets and networks, *The Age of the Network* and *The Team-Net Factor*. Finally, we encourage you to contact us directly with your virtual team story at: virtualteams@netage.com.

NOTES

Introduction   Coming Home

1. Jessica Lipnack and Jeffrey Stamps, *The TeamNet Factor: Bringing the Power of Boundary Crossing into the Heart of Your Organization* (New York: John Wiley & Sons, 1993).
2. Jessica Lipnack and Jeffrey Stamps, *The Age of the Network: Organizing Principles for the 21st Century* (New York: John Wiley & Sons, 1994).
3. Jessica Lipnack and Jeffrey Stamps, *Networking: The First Report and Directory* (New York: Doubleday, 1982).
4. Jessica Lipnack and Jeffrey Stamps, *The Networking Book: People Connecting with People* (New York: Viking Penguin, 1986).
5. Jeffrey Stamps, *Holonomy: A Human Systems Theory* (Seaside, CA: Intersystems Publications, 1980).
6. We organized the ten network principles in *Networking* in two groups of five: Structure (Holons, Levels, Decentralized, Fly-Eyed, and Polycephalous) and Process (Relationships, Fuzziness, Nodes and Links, Me and We, and Values).
7. The five networking principles in *The TeamNet Factor* and *The Age of the Network* are: Unifying Purpose, Independent Members, Voluntary Links, Multiple Leaders, and Integrated Levels. We represent all of these concepts in the nine principles of *Virtual Teams*.

Chapter I   Why Virtual Teams

1. According to the *Merriam Webster Collegiate Dictionary*, 10th ed., the correct way to spell this word is "collocated." Some people prefer "colocated." We went with the dictionary.
2. "Teams Become Commonplace in U.S. Companies," *The Wall Street Journal*, November 29, 1995, p. 1.
3. The source for annual global PC sales is Windows Internet Magazine World Wide Web site, www.winmag.com.
4. The source for annual global cellular phone sales is Action Cellular's World Wide Web site, www.snider.com.
5. An excellent source of information about the Internet is available from Matrix Information and Directory Services: www.matrix.org. Its careful

research makes valuable distinctions between computer networks, e-mail connections, number of computer servers, and the like. Since there is no universally accepted definition of "the Internet," it is impossible to precisely calculate its growth.

6. This is the subtitle of Ray Grenier and George Metes' book, *Enterprise Networking: Working Together Apart* (Bedford, MA: Digital Press, 1992).

7. For more on the relationship between proximity and collaboration, see Thomas J. Allen, *Managing the Flow of Technology: Technology Transfer and the Dissemination of Technological Information within the R&D Organization* (Cambridge, MA: MIT Press, 1977). Data are given in *The Age of the Network*, p. 47.

8. E. T. Hall, *The Hidden Dimension* (Garden City, NY: Doubleday, 1966).

9. For more on CERN, see its World Wide Web site: www.cern.ch.

10. Computer visionary Doug Engelbart, who in 1968 invented both the mouse and pull-down windows in computer programs, also designed his earliest systems in hypertext. For more information on Engelbart's work, see his World Wide Web site: www.bootstrap.org.

11. Virtuous is the term for positive feedback popularized by Peter Senge. See his book, *The Fifth Discipline: The Art and Practice of the Learning Organization* (New York: Doubleday/Currency, 1990).

12. Bernard DeKoven, *Connected Executives: A Strategic Communications Plan* (Palo Alto, CA: Institute for Better Meetings, 1990).

Chapter 2 Teaming from the Beginning

1. For more on the importance of the span of influence, see Reuben T. Harris, "Think Spans of Influence, Not Spans of Control," *The Tom Peters Group Update 1*, No. 2 (1991).

2. Alvin Toffler, *The Third Wave* (New York: William Morrow, 1990).

3. In our book, *The Age of the Network*, we describe the evolution of organization through these four eras on pages 12–13.

4. For an insightful and detailed look at the varieties of bureaucracy and their different paths of transition to networked organizations, see Raymond E. Miles and Charles C. Snow, *Fit, Failure & the Hall of Fame: How Companies Succeed or Fail* (New York: The Free Press, 1994).

5. Howard Rheingold, *The Virtual Community: Homesteading on the Electronic Frontier* (Reading, MA: Addison-Wesley, 1993).

6. Among the harbingers is NetResults, the agency spanning network of U.S. bureaucrats, that came together in 1993 as a result of the federal Reinventing Government program. See Lipnack and Stamps, *The Age of the Network*, pp. 127–133.

7. Most subsciences, specialized disciplines, and research concentrations have what is colloquially known as a "standard model." The standard model is the best current synthesis of available research, explanatory theory, and consensus thinking of the recognized leaders in a particular field. Of course, the model is always in motion and subject to reinterpretation with new information. By nature, the target of challenges, the model is also the productive source of new hypotheses and new ways to integrate existing information. For companies, their own "standard model" is what Peter Drucker calls their "theory of the business."

8. Richard A. Guzzo, Eduardo Salas, and Associates, *Team Effectiveness and Decision Making in Organizations* (San Francisco: Jossey-Bass, 1995), p. 115.

9. In systems parlance, this is the *sine qua non* characteristic of "nonsummativity," meaning that the whole is greater than the sum of the parts.

10. As the 1996 quote begins: " . . . teams share the foregoing characteristics with small groups, with one additional characteristic." Guzzo et al., *Team Effectiveness and Decision Making in Organizations,* p. 115.

11. An oft-quoted research definition of teams offers the three small group characteristics together with a task-oriented purpose: "Teams are distinguishable sets of two or more individuals who interact interdependently and adaptively to achieve specified, shared, and valued objectives." Guzzo et al., *Team Effectiveness and Decision Making in Organizations,* pp. 13 and 115.

12. CALS stands for Computer-assisted Acquisition and Logistic Support.

13. Robert Kraut and Carmen Egido, "Patterns of Contact and Communication in Scientific Research Collaboration," *Computer-Supported Cooperative Work,* Conference Proceedings (New York: Association for Computing Machinery, 1988).

14. Grenier and Metes, *Enterprise Networking.*

Chapter 3   The Power of Purpose

1. Robert Joines, "Eastman's Quality Journey: Chapter One" (presented at the Quest for Excellence Conference, Washington, DC, February 1994).

2. Lipnack and Stamps, *The Age of the Network,* pp. 52–58.

3. Ibid.

4. Lipnack and Stamps, *The Age of the Network,* p. 14.

5. E-mail from Yurij Wowczuk, May 15, 1996.

6. Inscription on Hopkins Memorial Steps, Williams College, Williamstown, MA.

7. Peter F. Drucker, "The Age of Social Transformation," *Atlantic Monthly* (November 1994), pp. 36–41.

Chapter 4 Through the Worm Hole

1. Technically, this high speed, high bandwidth connection is called a "switched T1 line."
2. Marshall McLuhan, *Understanding Media: The Extensions of Man* (New York: McGraw-Hill, 1964).
3. Lipnack and Stamps, *The Age of the Network*, p. 42.
4. Nicholas Negroponte, *Being Digital* (New York: Knopf, 1995).
5. For more information on the use of the word "matrix" see Matrix Information and Directory Services' World Wide Web site: www.matrix.org.
6. DeKoven, *Connected Executives*.

Chapter 5 Teaming with People

1. Tetra Pak was established in Lund, Sweden, in 1951 as a subsidiary of Akerlund and Rausing, formed in 1930.
2. Information about Tetra Pak was gleaned from a Web search, including www.westnet.se, a Swedish industry and trade Web site and "The News," Portugal's national newspaper.
3. Many details about the organization were provided in the translation of a chapter in Anders Högström, *Vinna Tillit* (Stockholm: Industrilitteratur, 1995).
4. Gary Hamel and C. K. Prahalad, "Core Competence of the Corporation," *Harvard Business Review,* May 1990, pp. 79–91.
5. Högström, *Vinna Tillit*, p. 14.
6. Arthur Koestler, *The Ghost in the Machine* (London: Hutchinson & Co., 1967). The holon has been part of our conceptual family for three decades now. Jeff found the hierarchy concept so pervasive in the systems literature, and the word holon so elegant in capturing the essence of the idea, that he titled his doctoral dissertation (and his 1980 book by the same name) *Holonomy,* which means "the study of holons." The holon "wholepart" was first among the ten principles of our first two books. Although we had sharpened the principles to five in *The TeamNet Factor* and *The Age of the Network*, in both books, we reintroduced the holon idea at the very end as part of the underlying systems framework supporting the network principles.
7. Herbert Simon, "The Architecture of Complexity," *Proceedings of the American Philosophical Society,* 1962.
8. Motorola 1995 Summary Annual Report, p. 10.
9. Luther P. Gerlach and Virginia Hine, *People, Power, Change: Movements of Social Transformation* (New York: Bobbs-Merrill, 1970).

10. Allen W. Johnson and Timothy Earle, *The Evolution of Human Societies: From Foraging Group to Agrarian State* (Palo Alto, CA: Stanford University Press, 1987), p. 52.

11. This is known as "metonomy" in that branch of cognitive science that looks at thinking through the categories (mental models) we use.

12. Lipnack and Stamps, *The Age of the Network*, p. 84, and *The TeamNet Factor*, pp. 47–49.

13. Glenn M. Parker, *Team Players and Teamwork: The New Competitive Business Strategy* (San Francisco: Jossey-Bass, 1991), p. 53.

14. Lipnack and Stamps, *The Age of the Network*, p. 85.

15. Johnson and Earle, *The Evolution of Human Societies*, p. 320.

16. Lipnack and Stamps, *The TeamNet Factor*, p. 13.

17. Research has repeatedly demonstrated the inverted "U"–shaped relationship between size and performance. Paul S. Goodman and Associates, *Designing Effective Work Groups* (San Francisco: Jossey-Bass, 1986), p. 16.

18. Robert Reich, "Entrepreneurship Reconsidered: The Team as Hero," *Harvard Business Review* (May-June 1987).

Chapter 6   It's All in the Doing

1. James Grier Miller, *Living Systems* (New York: McGraw-Hill, 1978).

2. Lipnack and Stamps, *The Age of the Network*, p. 231.

3. Gilbert Amelio and William Simon, *Profit From Experience* (New York: Van Nostrand Reinhold, 1996).

4. Ludwig von Bertalanffy, *General Systems Theory: Foundations, Development, Applications* (rev. ed.) (New York: George Braziller, 1968).

5. Lipnack and Stamps, *The TeamNet Factor*, pp. 221–223.

6. Ibid., Chapters 8–10.

7. Senge, *The Fifth Discipline*.

8. Technically, "slowing" is negative feedback, "growing" is positive feedback.

9. Jessica Lipnack and Jeffrey Stamps, "The Virtual Water Cooler: Solving the Distance Problem in Networks," *Firm Connections 1*, No. 2 (May-June 1993).

10. Dean W. Tjosvold and Mary M. Tjosvold, *Leading the Team Organization: How to Create an Enduring Competitive Advantage* (New York: Macmillan, 1991); Dean W. Tjosvold, *Working Together to Get Things Done: Managing for Organizational Productivity* (Lexington, MA: D.C. Heath, 1986).

11. Tjosvold, *Working Together to Get Things Done*, pp. 32–33.

12. We intentionally use codependent, the popular word from psychology, here. Codependent relationships are not healthy. Used in relationship to the world of work, the word means that one person can win only if someone else loses.

13. *MacWEEK,* June 24, 1996, p. 68.

14. Lipnack and Stamps, *The Age of the Network,* pp. 16–17, and *The Team-Net Factor,* p. 11; Adam M. Brandenberger and Barry J. Nalebuff, *Co-opetition: A Revolutionary Mindset That Combines Competition and Cooperation: A Game Theory Strategy That's Changing the Game of Business* (New York: Doubleday, 1996).

15. For more on this, see Lipnack and Stamps, *The Age of the Network,* Chapter 8.

Chapter 7 *Virtual Place*

1. Sun is an acronym for Stanford University Network. Three of Sun's founders were Stanford alumni. Their mission was to build a company "to provide 'open' desktop computers at one-tenth the cost of existing systems."

2. Ryan Bernard, *Corporate Intranet: Create and Manage an Internal Web for Your Organization* (New York: John Wiley & Sons, 1996), p. 134.

3. Joshua Meyrowitz, *No Sense of Place: The Impact of Electronic Media on Social Behavior* (New York: Oxford University Press, 1985).

4. Ibid.

5. Ibid.

6. The Education Development Center was founded in 1958 by a group of MIT scientists to develop a new curriculum for high school physics. Today it is an international research and development organization "dedicated to building talent and know-how for human advancement."

7. Grenier and Metes, *Going Virtual.*

8. Lipnack and Stamps, *The TeamNet Factor,* pp. 31–34.

9. For more information on Lynx, start at either the Lynx Enhanced Pages at http://www.nyu.edu/pages/wsn/subir/lynx.html or Al Gilman's FAQ (frequently asked questions) at http://www.access.digex.net/~asgilman/lynx/FAQ.

10. Kathleen K. Mall and Sirkka L. Jarvenpaa, "Learning to Work in Distributed Global Teams." This paper is available online at the World Wide Web site: uts.cc.utexas.edu/~bgac313/hicss.html.

11. Robert K. Greenleaf, *Servant Leadership: A Journey into the Nature of Legitimate Power and Greatness* (Mahwah, NJ: Paulist Press, 1977).

12. TCP/IP means Transmission Control Protocol/Internet Protocol.

13. Lipnack and Stamps, *The TeamNet Factor,* p. 331.
14. Bernard, *Corporate Intranet,* p. 136.

Chapter 8 Working Smart

1. In planning sessions with teams, we peel back purpose from goals to results to tasks, the activities necessary to get from here (goals) to there (results). Lipnack and Stamps, *The Age of the Network,* p. 166, and *The TeamNet Factor,* p. 255.
2. Lipnack and Stamps, *The TeamNet Factor,* Chapter 10.
3. TeamFlow is produced by CFM, Inc., 60 The Great Road, P.O. Box 353, Bedford, MA 01730–0353; phone: 617/275–5258; e-mail: VTInfo@Team-Flow.com; and World Wide Web site: www.teamflow.com.
4. For readers of our previous books, the Virtual Team Pocket Tool is the next iteration of the model we have been developing since 1979.
5. Guzzo, Salas, and Associates, *Team Effectiveness and Decision Making in Organizations.*

Chapter 9 Virtual Values

1. For an extended study of Harry Brown and EBC Industries, see Lipnack and Stamps, *The TeamNet Factor,* pp. 137–139, and *The Age of the Network,* pp. 79–85.
2. The Deming quote is from the foreword to John O. Whitney, *The Trust Factor: Liberating Profits and Restoring Corporate Vitality* (New York: McGraw-Hill, 1994), p. viii.
3. Ibid., pp. 18–19.
4. Lee Sproull quote is from "Virtual Teams," by Beverly Geber, *Training,* April 1995. See also Sproull and Sara Kiesler, *Connections: New Ways of Working in the Networked Organization* (Cambridge, MA: MIT Press, 1993).
5. Gates is referring to the use of keyboard symbols to depict people's feelings. Called "emoticons" (the computer-lingo contraction for emotional icons), they must be read horizontally to make sense and include such symbols as the "smiley face" that Gates references: :-) For a wink: ;-) For sadness: :-(
6. John Case, *Open Book Management* (New York: HarperCollins, 1996) and John Schuster, *The Power of Open Book Management* (New York: John Wiley & Sons, 1996).
7. James S. Coleman, "Social Capital in the Creation of Human Capital," *American Journal of Sociology* (1988 Supplement), S98.

8. Ibid.

9. Robert D. Putnam, "Bowling Alone: America's Declining Social Capital," *Journal of Democracy 6,* No. 1 (January 1995), pp. 65–78, and "Bowling Alone, Revisited," *The Responsive Community* (Spring 1995), pp. 13–33.

10. Robert D. Putnam, *Making Democracy Work: Civic Traditions in Modern Italy* (Princeton, NJ: Princeton University Press, 1993). See also Lipnack and Stamps, *The Age of the Network,* Chapter 8, for a broader discussion.

11. Tariq Banuri, Goran Hyden, Calestous Juma, and Marcia Rivera, "Sustainable Human Development: From Concept to Operation: A Guide for the Practitioner" (discussion paper, United Nations Development Programme, 1994), p. 21.

12. For more information on Smart Valley, see its World Wide Web page: www.svi.org.

13. For more information on Joint Venture Silicon Valley Network, see its World Wide Web page: www.jointventure.org.

14. AnnaLee Saxenian, *Regional Advantage: Culture and Competition in Silicon Valley and Route 128* (Cambridge, MA: Harvard University Press, 1994) and Lipnack and Stamps, *The Age of the Network,* Chapter 8.

15. Ibid.

16. See *Grassroots Leaders in the New Economy: How Civic Entrepreneurs Are Building Prosperous Communities* by Douglas Henton, John Melville, and Kimberly Walesh (San Francisco: Jossey-Bass, 1997). Henton and his colleagues did the original research that led to the formation of Joint Venture Silicon Valley Network and have served as its principal consultants since its inception. For more information, contact them at: Collaborative Economics, 350 Cambridge Avenue, Suite 200, Palo Alto, CA 94306; phone: 415/614–0230; fax: 415/614–0240; e-mail: CoEcon@aol.com.

17. Our colleague, Charles Snow, Professor of Business Administration at the Pennsylvania State University, believes the most successful networks are in fact "cellular organizations."

18. Banuri, Hyden, Juma, and Rivera, "Sustainable Human Development," p. 19.

19. The Great Plains Partnership (GPP) was initiated by then Governor Mike Hayden of Kansas, under the auspices of the Western Governors' Association, which maintains an active role in the group. For more information about GPP, see its home page: http://rrbin.cfa.org/rrbin/gpp/gpphome.html.

20. "Webmaster" is the term that refers to the person who designs and maintains a World Wide Web site.

21. *The Wall Street Journal,* June 13, 1996, B1.

About the Authors

Jessica Lipnack and **Jeffrey Stamps, Ph.D.,** are principals of The Networking Institute, Inc. (TNI), a consulting company they founded in 1982. TNI helps organizations develop networks and virtual teams. Clients include Apple Computer, AT&T Universal Card Services, Bank-Boston, Digital Equipment Corporation, Hewlett-Packard, Intel, NCR, Steelcase, Rodale Press, the Massachusetts Teachers Association, and The United Nations.

Lipnack and Stamps are leading experts in networked organizations. Quoted often in the press, they lecture extensively around the world. They have appeared on many radio shows and have written numerous articles for major publications. They also have conducted scores of in-house workshops that have launched networks of all kinds within companies, nonprofit organizations, government agencies, major religious denominations, and educational institutions.

Virtual Teams is their fifth book on this topic.

For more information:

> The Networking Institute, Inc.
> 505 Waltham Street
> West Newton, MA 02165
> USA
> Phone: (617) 965-3340
> Fax: (617) 965-2341
> E-mail: virtualteams@netage.com
> Web site: www.netage.com

Index

255